Agency, Democracy, and Nature

Agency, Democracy, and Nature

The U.S. Environmental Movement from a Critical Theory Perspective

Robert J. Brulle

The MIT Press
Cambridge, Massachusetts
London, England

Set in Sabon by The MIT Press.
Printed and bound in the United States of America.

Library of Congress Cataloging-in-Publication Data

Brulle, Robert J.
Agency, democracy, and nature : the U.S. environmental movement from a critical theory perspective / Robert J. Brulle.
p.   cm.
Includes bibliographical references and index.
ISBN 0-262-52281-0 (pb) — ISBN 0-262-02480-2 (hc)
1. Environmentalism—United States—History. I. Title.
GE197.B77 2000
333.7'2'0973—dc21   00-028264

to Joe and Tim, who will inherit the results of our actions

Contents

Acknowledgements

While this book has one author's name on the cover, it is, in fact, the result of an extended conversation. There are many individuals who made substantial contributions to this effort. First, I would like to offer a sincere thanks to Richard Harvey Brown. His selfless giving of knowledge, time, and friendship were vital to my development as a scholar and person. Special thanks also goes to my dissertation committee at George Washington University, including Ruth Wallace, Steve Tuch, Pat Lengermann, and especially Thomas Dietz. Tom's interest and detailed involvement in this research contributed heavily to its completion.

In the wider intellectual community, I was fortunate to meet Riley Dunlap early on in my academic career. His continued interest, support, and friendship have made my entry into the community of environmental sociologist an enjoyable and enhancing experience. I also benefited greatly from my conversations with a number of scholars, including Tom Burns, Beth Schaefer Caniglia, Carlo Jaeger, Craig Jenkins, Gene Rosa, Donald Snow, Paul Stern, and Tom Webler. In addition, the anonymous reviewers were very useful in helping me to focus and clarify my arguments.

While preparing the manuscript, I had the opportunity to speak at length with several individuals involved in environmental politics. These individuals, including Ron Arnold, David Brower, Barry Commoner, Dave Foreman, Lois Gibbs, Gary Snyder, and Terry Swearingen all provided me with a number of useful insights. I am grateful to them for taking the time to talk with me.

The staff at The MIT Press also deserves special recognition. First, I would like to thank the manuscript editor for his outstanding efforts, which clarified and strengthened my arguments. Second, my acquisition editor, Clay Morgan, was a continued source of support.

Finally, Miranda Spencer provided me with much positive encouragement and support while I completed this book, for which I am very grateful. Her skillful editing greatly enhanced the readability of the book. In addition, my two sons, Joe and Tim Brulle, to who this book is dedicated, had to put up with my continued hermit-like existence while completing my Ph.D. and then finishing the book. Their patience with me is appreciated.

Agency, Democracy, and Nature

1
Ecological Degradation and Social Change

The environmental effort has largely failed.
—Barry Commoner (1991: 63)

The global ecocrisis continues to worsen despite people's efforts to respond.
—Max Oelschlaeger (1994: 4)

The task of developing an ecologically sustainable society is a major challenge facing current social institutions. However, the results achieved so far make this imperative appear to be only a utopian fantasy, fast receding from our grasp. Since the first Earth Day (1970), a number of actions have been taken in the United States to deal with environmental degradation. Although there has been incremental progress in reversing some of the worst forms of visible pollution, it pales in comparison to the changes that are needed.

We are losing the struggle to reverse ecological degradation. The evidence for this is beyond question.

First, we are exceeding the capacity of the natural environment to absorb our waste products. This is evident in the spread of air and water pollution, the disposal of toxic and other wastes, and the spread of pesticides and hormone disrupters. More than 100 million Americans live in metropolitan areas where the U.S. Environmental Protection Agency classifies the air quality as unhealthy. Acid rain continues to fall on the eastern states. Nearly 40 percent of U.S. rivers and lakes are not fit for drinking, fishing, or swimming. Virtually all the fish in the Great Lakes are unfit for human consumption owing to the presence of persistent carcinogenic pollutants, such as mercury, polychlorinated biphenyls (PCBs), dioxin, and DDT. Despite more than 15 years of effort since the passage of the first Superfund bill,[1] by 1995 treatment had been completed at only 25 percent of the so-called National Priority cleanup sites.

Second, we are depleting a number of resources that are necessary for human life. The global cycles of carbon, water, and nitrogen have been altered by human activities, leading to severe ecological problems, including shortages of potable groundwater, alteration of the food chain, and declining soil productivity. We are also exhausting the renewable resources of the planet. Nearly two-thirds of ocean fisheries are overexploited or near their limit of exploitation (Vitousek et al. 1997). Topsoil loss continues, exceeding its rate of replacement by 15 to 30 times, and aquifers are being depleted around the world (Abernethy 1998: 106).

Third, we are transforming increasing portions of the natural world to serve human needs. In the process, we are destroying the ecosystems on which human survival is based. Humans now use 40 percent of the planet's capacity to support life for their own activities. We are destroying increasing proportions of plant and animal life. Deforestation is accelerating. More than half of the forests are gone, and we continue to lose 16 million hectares annually (Abramovitz 1998: 22). According to the ecologist Norman Myers (1996: 133), "not only do we appear set to lose most if not virtually all tropical forests, but there is progressive depletion of tropical coral reefs, wetland, estuaries, and other biotopes."

The resultant irreversible global loss of biodiversity is staggering. As documented by the International Union for the Conservation of Nature, 25 percent of mammalian species, 11 percent of bird species, 20 percent of reptiles, 25 percent of amphibians, and 34 percent of fishes are threatened with extinction (IUCN 1996). In addition, more than 12 percent of tree species face the threat of extinction (IUCN 1997). As these species disappear, so too does the integrity and functioning of the ecosystem that sustains human existence. It is now certain that humans have disrupted evolutionary and ecological processes on a global scale.

Finally, human activities are disrupting the global climate. The Second Assessment by the Intergovernmental Panel on Climate Change (IPCC 1995) shows that climate change is underway and is increasing. Already there has been an increase in the global average temperature and a rise in sea level. These trends are projected to increase. The consequences of these changes will be substantial, if not devastating for many species. Severe losses in biodiversity are expected as rapid climate shifts restrict native habitats.

Sea-level rise will eliminate many salt-water marshes, mangrove forests, and coral reefs, with a concomitant loss of marine ecosystems and species.

Human activities will also be severely impacted. Coastal economic activities will be disrupted and will have to be relocated. Large-scale public works projects will be needed to save many coastal cities. In poorer countries, where these forms of adjustment are unaffordable, the consequences will be even more severe than elsewhere. For example, a one-meter rise in sea level would result in an 18 percent loss in the land area of Bangladesh, thus creating extraordinary restrictions in available food supplies, potable water, and living space. Mass migrations and extensive human suffering should be expected. Human health will also suffer. Increases in heat stress, air pollution, vector-borne infectious diseases, populations of pest species, and shifts in food production will occur. Global warming will have "pervasive adverse impacts on human health and result in significant loss of life" (International Physicians 1997).

The types of ecological destruction appear to go on indefinitely, and new problems constantly arise. It is as if Western society has deliberately set out to destroy the integrity of the ecosystem. There is no sign that these conditions are abating. Instead, they appear to be accelerating. Every year since 1984, the Worldwatch Institute has issued a report on the ecological condition of the world. In reviewing these trends, Brown (1991: 154) concluded that "each of the vital signs shows continuing deterioration from year to year."

The overall results of this process are projected to be irreversible degradation of many ecosystems, extinction of many species of plants and animals, and a severe cumulative adverse effect—not only for the other beings with which we share the planet but also for us. Human health is already suffering from ecological degradation. According to Chivian et al. (1993: 2):

In the United States alone, 100 million people breathe levels of ozone dangerous to health at various times of the year. Acid rain is causing damage to forest and aquatic ecosystems on every continent. Depletion of the stratospheric ozone layer, which protects us from harmful solar ultraviolet rays, and changes in global climate due to the emission of greenhouse gases into the atmosphere have the potential to affect virtually every human being on the planet. . . . We are altering the basic physiology of the planet. . . . Tropical forest destruction, soil erosion, the withdrawal of water from the hydrologic cycle, and toxic pollution are causing ecological changes which threaten the quality of life and indeed the survival of humanity.

For humans, the cumulative result of ecological degradation is an overall decline in the quality of life and in life expectancy. Meadows et al. (1992: xv–xvi) argue that the ecological carrying capacity of the planet has been exceeded as a result of human activities, and that "without significant reductions in material and energy flows, there will be in the coming decades an uncontrolled decline in per capita food output, energy use, and industrial production." What these scarcities and privations will mean for human society can only be projected. Severe ecological degradation may lead to social disruptions and human conflict. Some scholars have concluded that "scarcities of renewable resources are already contributing to violent conflicts in many parts of the developing world" (Homer-Dixon et al. 1993: 38). These conflicts are projected to increase as resource scarcity increases (Dobkowski and Wallimann 1998: 198).

Are all the projections correct? Certainly not. However, the possibility that they are all *incorrect* is zero. Virtually all the projected trends point toward significant and irreversible problems due to ecological degradation. These projections must be taken seriously if we are to formulate actions to mitigate their consequences.

And we are failing to take meaningful action. Immense changes are required just to stop (much less reverse) ecological degradation. For example, to stabilize greenhouse-gas emissions at twice pre-industrial levels would require a 50 percent reduction in global CO_2 emissions. Just to stabilize the atmospheric concentration of greenhouse gases at current levels would require reducing worldwide CO_2 emissions by a factor of 5 or 6 (Holdren 1997). This would entail major transformations in our energy and transportation infrastructures. We have not addressed this problem in any meaningful way. In fact, we have difficulty in developing and enforcing minimum fuel-efficiency standards for automobiles. Even if these changes were implemented, the addition of more automobiles would likely offset any net reduction in CO_2 emissions.

Our attempts to deal with environmental degradation are, so far, basically feeble and symbolic actions without any measurable results (Trainer 1998: 85). The cumulative consequences of our society's efforts to address these problems remains negligible. Surveying our environmental efforts, Ponting (1991: 400) has concluded that "compared with the scale of the problems, many measures are little more than cosmetic."

Social Change and Ecological Sustainability

To move beyond the current impasse, we need to develop and implement major and meaningful measures that would reverse the destruction of the Earth's ecosystem. The major challenge faced by human society is "to anticipate or recognize at what point the environment is being badly degraded by the demands placed upon it and to find the political, economic and social means to respond accordingly" (Ponting 1991: 407).

An analysis of current efforts to deal with ecological degradation can be found in the National Research Council's report *Global Environmental Change: Understanding the Human Dimensions* (Stern, Young, and Druckman 1992). This report argues that most recent and proposed cures focus primarily on the *proximate* causes of environmental degradation: production of toxins, emission of greenhouse gases, and pollution of water. These forms of analysis overlook the social interactions that underlie environmental degradation (ibid.: 75). In order to lead to meaningful measures to mitigate ecological degradation, this report argues, an analysis of ecological degradation must be based on an understanding of the social processes of the society in which these problems originate (ibid.: 17).

Environmental problems are fundamentally based on how human society is organized. Accordingly, social change is required for their resolution. Social change does not come about quickly or easily. The experience of the U.S. environmental movement since the first Earth Day shows the great difficulty of reorganizing the social order so it can function without destroying its ecological foundation.

The current social order is the cumulative result of the modern social project. More than 500 years of human activity has been directed toward its creation. The transition to an ecologically sustainable society involves changes at least as great as those that accompanied the transformation from an agricultural to an industrial society. To reverse this cumulative history of destruction of the natural world will certainly take much more time and effort than was previously imagined.

Large-scale social change occurs over a long time period, through the gradual accumulation of a myriad of changes. The direction of this process is not intentionally determined by particular individuals or groups. No one set out to create capitalism, the nation state, or an industrial society. It took several centuries for the current social order to develop. However, ecological degradation continues to accelerate. As a consequence, ecological problems must be dealt with in much shorter time frames than can be anticipated for slow, gradual, or unintended social change. Thus, we must focus on rapidly developing a society that is ecologically sustainable. We cannot leave this up to vague, indeterminate, undirected social change.

It does not appear that the requisite kind of social change will be supported by the existing power structure. Change is normally resisted by powerful elites, and there is no special exemption for societies experiencing ecological degradation. "One lesson to be learned from the history of past civilizations," Gowdy (1998: 74) writes, "is that political pressure from elites bent on preserving their power at all cost will inevitably result in perpetuating unsustainable systems until it is too late, that is, until environmental degradation leads to social disintegration."[2]

This situation defines a social project that has never been successfully completed and for which there are no historical guidelines. Yet the need for action is clear. Commoner (1971: 300) has offered the following comments:

> . . . none of us—singly or sitting in committee—can possibly blueprint a specific "plan" for resolving the environmental crisis. . . . The world is being carried to the brink of ecological disaster . . . by the phalanx of powerful economic, political, and social forces that constitute the march of history. Anyone who proposes to cure the environmental crisis undertakes thereby to change the course of history. . . . But this is a competence reserved to history itself, for sweeping social change can be designed only in the workshop of rational, informed, collective social action. That we must act is now clear. The question which we face is how.

Attempts to deal with global ecological deterioration have spawned a virtual environmental establishment—innumerable international, national, and local government agencies, scientific research programs, university programs, think tanks, foundations, and thousands of social-movement organizations. Yet, despite all these well-intentioned and hard-fought efforts, the scope and the severity of ecological problems continue to increase. We are in a condition that Oelschlaeger (1994: 4) has called the Paradox of Environmentalism: "The global ecocrisis continues to worsen despite people's efforts to respond."

Although various social-movement organizations have developed numerous alternative narratives, none of them has been able to develop a significant challenge to the dominant belief system of Western society. As Goldblatt (1996: 202) notes, "contemporary ecology movements have yet to find the public idiom and institutional format in which that argument, in the face of structural resistance and public apathy, becomes utterly compelling."

In addition, many of the proposed alternatives are badly flawed. Though there have been many specific proposals for bringing about an ecologically sustainable society, the means by which a transition to a social order that could implement these changes might be accomplished remain vague. There is no plausible theory of transition to a sustainable society. Dobson (1993: 192) has noted that "the issue of social change is undertheorized in green politics." Commoner (1991: 63) has written that "somebody must say: The environmental effort has largely failed, and here is what we ought to do to rebuild it."

Faced with an unprecedented and serious situation, we are "as a culture searching for how to respond to our intensifying sense of ecological crisis" (Kearns 1996: 64). If an ecologically sustainable society is to be created, social learning must be rapidly expanded, resistance must be overcome, and intentionally directed social change must be accelerated. This is a difficult theoretical and practical task, but it must be undertaken if we are to avoid the extraordinary suffering and misery that will be inflicted on all of the Earth's living beings if the projected levels of ecological disruption occur (Dobkowski and Wallimann 1998).

Social Science and Ecological Degradation

To develop and carry out a plan of action to deal with the process of ecological degradation will required social-science research that can foster social change. What is needed is for social scientists to "increasingly engage in a kind of scientific praxis that is explicitly directed toward the preservation of life, at home and elsewhere in our global society" (Dobkowski and Wallimann 1998: 4).

There are many schools of thought within social science, from empiricism to postmodernism. Moreover, there is no consensus on which approach should be used to examine social phenomena, much less on how ecological degradation should be dealt with. After reviewing the adequacy of current social theory to address this problem, Goldblatt (1996: 203) drew these conclusions:

> We know that environmental degradation is dangerous. We know that we cannot go on as before. But how to go on, how to live individually and collectively, how to make the transition soon and how to persuade the intransigent, the selfish, the powerful, and the uninterested? These are the questions that neither classical socialism nor contemporary social theory have provided sufficient intellectual or moral resources to answer.

One important and promising approach has been developed by the German sociologist Ulrich Beck, who maintains that we need to shift the focus of our understanding and research on the process of ecological degradation from the physical and natural sciences to an analysis of the social origins of ecological degradation. In Beck's words (1995: 16), "from a scientifically inspired policy of revealing horror scenarios to a social science based redirection of accountability."

To address the social origins of ecological degradation would require the development of a social theory that could show how the processes of ecological degradation are based on cultural beliefs and social institutions, and thus inform political practice (Beck 1995: 140–141). This would then enable social movements to make the social production of risks visible, and thus put the problems of a risk society on the public agenda (Beck 1992a: 116; Beck 1995: 32). It is hoped that this process would result in the creation of a "green" public sphere, which would enable us to debate and make democratic decisions about how we should act to create an ecologically sustainable society (Torgerson 1999: 160).

Critical Theory and Social Change

A cogent and powerful argument has been developed for using critical theory in the development of a social science that would unite theory with practice, thus enabling us to develop the cognitive, moral, and aesthetic cultural resources needed to address ecological degradation.[3] Critical theory, as originally defined by Marx (1843: 209), has the task of "the self-clarification of the struggles and wishes of the age." Critical theory aims at facilitating the creation of a social order founded upon the reasoned acceptance of social organization freely arrived at in public dialogue. With will and knowledge, the human species is seen as able to create its own nature and future through self-critical renewal of the social order (Horkheimer 1937: 245). By enhancing social learning, critical theory can assist in the reformulation of the social order to create an ecologically sustainable society.

In this perspective, social order is seen as being formed through consciously initiated actions and also through aggregate dynamics. The aggregate dynamics take the form of the "unplanned consequences, unrecognized interdependencies, [and] uncomprehended system dynamics [that] hold sway over our lives like a second nature" (McCarthy 1985: 50). Social institutions appear as objective and fixed realities, not as results of human construction. The nature of these social dynamics is generally outside our everyday understanding. This aspect of social life was captured by Foucault: "People know what they do; they frequently know why they do what they do; but what they don't know is what what they do does." (Dreyfus and Rabinow 1982: 187)

To enhance the capacity for social learning, critical theory seeks to foster an understanding of the current social organization, including previously uncomprehended social dynamics. The goal is to enable us to gain a greater aspect of control over our own futures. Specifically, critical theory attempts to increase the extent of rational control of social organization by bringing the processes that create and maintain social institutions into our collective vision, so that the "objective and natural" form of those social institutions ceases to exist. Instead, the institutions are seen as human constructions that can be changed (Lukacs 1968: 17–18). Thus, critical theory seeks to expand our social horizons. It strives to foster an examination of our social behavior and beliefs, to develop the recognition that the existing state of affairs does not exhaust all possibilities, and to offer suggestions for the conscious development of more effective and morally sensitive actions.

The role of the critical social theorist starts with accepting the concerns and issues of his or her particular social group as a research problematic. The theorist then uses his or her specialized training and skills to "make coherent the principles and the problems raised by the group in their practical activity" (Gramsci 1971: 330). Thus, intellectual activities are not scholastic activities played out among an elite group of literati, but rather the practical development of a life practice.

It is important to note that in creating this theoretical perspective the social theorist does not occupy a uniquely objective or privileged position in viewing the social order. There is no universal or archimedian position for gaining knowledge about society. Instead, the social theorist participates in a particular social situation and views reality through a specific historical framework. His research focus comes from participation in social movements and seeks to address his concerns and questions as he carries on his practical activities.

Critical theory thus defines a particular type of research program that can unify theory and practice. The focus of the research is on the concerns and questions of a group in society. The goals of the research are to form a cogent model of the social order that can stimulate the particular social group's self-reflective capacity and to facilitate public, democratic deliberation over what actions should be taken. These goals are accomplished by forming a coherent worldview and effectively communicating this view in the public arena. This worldview then can help foster authentic insights into the situation of the group, and can serve as the basis for public deliberation about what actions should be taken (Calhoun 1995; Bronner 1994; Turner 1992; Dryzek 1995: 99). Once the unconscious formation of social order has been brought into conscious reflection, the human social order can take control over it. In the words of Mills (1959: 6), this knowledge might "enable us to grasp history and biography and the relations between the two within society" and then enable us to act on the basis of this knowledge. Social research based on critical theory thus seeks to make social development a conscious and rational act that citizens control:

... it means that you put the responsibility for decisions on the shoulders of those who anyhow will suffer their consequences, and that at the same time you stimulate the participants who have to make up their minds in practical discourse to look around for information and ideas that can shed light on their situation— which can clarify their understanding of themselves. (Habermas 1986b: 207)

The potential of critical theory has not been realized in recent years. Instead, critical theory has been subjected to a "deadening scholasticism" (Bronner 1994: 322). Few applications of critical theory have used empirical research to examine questions regarding the public interest. Instead of focusing on integrating theory and practice, critical theory has engaged in an exegesis of the theoretical and philosophical concepts that constitute it.

As a result, critical theory is primarily an activity in the academic world, with little or no influence outside of that cloistered environment (Bronner 1994: 322; Turner 1992: 165).[4] Bronner (1994: 322) describes this situation as follows: "A metaphysical veil has fallen over critical theory. . . . A new identification with the disempowered can help lift this veil. Providing critical theory with a new positive direction of this sort, however, is possible only by reaffirming its forgotten materialist component and reinvesting it with a practical interest in public affairs."

More specifically, critical theory has failed to meaningfully engage the problem of environmental degradation. Its major focus has been on the development of a more comprehensive notion of a rational society. With this focus, it has not developed an adequate analysis of the material conditions for its existence, or a strategy of transition to a society that would be both democratic and ecologically sustainable.

This strategy of transition is a crucial project for individuals and social movements trying to formulate a response to the growing problem of ecological degradation. Yet, by demanding strategies for action, "green" politics challenges critical theory and points to its current inadequacies (Dobson 1993: 209). In Dobson's (1996: 298) opinion, "critical theory might provide a historical and material analysis of the relationship between human beings and the natural world, together, perhaps with a non-Utopian resolution of the contemporary difficulties with this relationship." Critical theory may thus hold the potential to create a new metanarrative that will be able to "overcome the environmental dilemma" (Killingsworth and Palmer 1992: 130).

A strategy for action based on critical theory "must propose ways of rendering anonymous institutions and their administrative sub-systems more accountable" (Bronner 1994: 348). This means, at the least, that research on issues of public concern is needed. Such empirical research would allow the testing of many of the hypotheses regarding Western society that have been developed by critical theorists. Since theoretical research and empirical research are mutually informative, this type of process would also enhance the further development of this theoretical perspective.

The work of today's premier critical theorist, Jürgen Habermas, forms the primary theoretical perspective of this book. Habermas's work is an honest and thoroughgoing attempt to address our current situation and problems. His diagnosis of modernity has spawned many fruitful efforts to spell out the historical and empirical mechanisms of the expansion of systems logic into the lifeworld and to create the intellectual foundations of a just and democratic political practice.

As I have noted, critical theory has some well-recognized shortfalls in regard to empirical research and political action. However, the purpose of this book is not to engage in further theoretical exegesis but to address these shortcomings with reference to the process of ecological degradation.[5]

Habermas's work provides one of the most comprehensive and cogent approaches to the study of the current social order. In view of the perceived inadequacy of our current cultural and social systems to solve the problems we are faced with, his project offers at least an attempt at an answer of how we can better organize our society. Throughout this text, a deliberate attempt is made to work within the broad framework that Habermas has developed. Although many different sociological approaches are used in carrying out this research, the main outlines of both the theoretical analysis and the research questions remain within the spirit, if not the letter, of his approach to social theory.

The Plan of the Book

We need to find new ways of dealing with the ongoing ecological degradation. Though there have been many well-intentioned efforts historically, they have not proved adequate to our current situation. We have to make a quantum increase in the rate of social change toward a sustainable society if we are to avert the worst of the projected consequences. The environmental social movement[6] in the United States is a key component in fostering social change.

In chapter 2 I review the existing research on the relationships between social dynamics and the natural environment, combining the various approaches to give an overview of the literature. I then build a basis for the use of critical theory to expand the examination of the social causes of ecological degradation. A review of Habermas's theory of communicative action sets the basis for the use of critical theory to develop an expanded model of the relationship between social dynamics and ecological degradation in chapter 3. I then develop a theoretical analysis that shows the vital role a democratic public dialogue plays in the renewal of social institutions to deal with the process of ecological degradation.

In chapter 4, through selective reading of the literature on frame analysis, resource mobilization, and historical sociology, I develop a framework within which to examine U.S. environmental organizations and to evaluate their capabilities and their limitations with respect to the creation of a democratic and sustainable society. In chapter 5 I begin an empirical examination of the U.S. environmental organizations by focusing on size and level of resource mobilization.

In chapters 6–9 I examine in detail the various groups that are concerned with how U.S. society should relate to its natural environment. Each chapter focuses on a specific community that is based on a particular worldview. This worldview is identified through a hermeneutic analysis of the historical and philosophical literature that created and defines the worldview. The instantiation of this worldview into movement organizations is then demonstrated through reference to examination of the founding documents of several paradigmatic environmental organizations in each community. A summary of the goals, strategies for social change, internal structure, and funding sources for each community is provided. A critical examination of each particular worldview is then conducted to determine whether or not it limits the range of responses to the problem of ecological degradation. This is followed by an analysis of the links between the worldview and organizational practices, and how this fosters or limits communication within the organization. By performing this analysis, this book seeks to provide an examination of how environmental movement organizations can contribute to the creation of a democratic and ecologically sustainable society.

In chapter 10 I summarize the results of the research across all the environmental communities, detailing their historical development and their distribution among the various discourses by membership, number of organizations, and funding levels. I also summarize the relationship between discourse and organizational practices. In chapter 11 I describe the practical implications of this research and suggest directions that the environmental movement can take to reform its organizational practices.

Although the environmental movement has the potential to foster the development of a sustainable society, there is no certainty that it will accomplish that task. The point of my analysis is to examine the practices of various organizations and to identify social processes that can assist in developing a democratic and sustainable society. To the extent that this type of analysis can expand our self-understanding, it can help us in renegotiating our social order and in choosing our future. If successful, this type of theoretical program could yield attempts to realize the utopian content of a society through trial and error in a "logic of justified hope and controlled experiment" (Habermas 1971: 283–284).

2

Communicative Action and the Social Order

The nonrenewable resources of our natural environment and the symbolic structures of our lifeworld—both the historically developed and the specifically modern forms of life—need protection. But they can be protected only if we know what
is threatening the lifeworld.
—Jürgen Habermas (1989b: 44)

In April of 1970, in the single largest demonstration in U.S. history, more
than 20 million people participated in the first Earth Day. Throughout the
country there were numerous demonstrations, marches, and speeches.
Several actions to counteract ecological degradation were advocated.
However, there was little consensus as to what was causing our ecological
problems or what we should do about them. Reviewing the speeches of
Earth Week, Commoner (1971: 6–10) noted that numerous "causes" of
ecological degradation were cited, including rising population, affluence,
man's innate selfishness and greed, religious belief systems, technological
development, political structures, and a capitalist economic system. This
led Commoner to attempt to "find some meaning in the welter of contradictory advice" (ibid.: 10). Commoner's conclusion was as follows: "The
confusion of Earth Week was a sign that the situation was so complex and
ambiguous that people could read into it whatever conclusion their own
beliefs—about human nature, economics, and politics—suggested. Like a
Rorschach ink blot, Earth Week mirrored personal convictions more than
objective knowledge."

Since 1970 there has been a virtual flood of literature on the causes of
and the cures for ecological degradation. The debate over the causes of ecological degradation that was evidenced in the speeches of the first Earth
Day continues today as one of the most contentious and divisive public

issues. One reason for this contentious debate is that its outcome has practical consequences. Every model of the causes of environmental degradation defines, implicitly or explicitly, the solution to this problem, and consequently defines the actions we should take. If ecological degradation is caused by expanding human populations, we need birth control; if it is caused by excessive consumption, we need to reduce our use of material goods; if it is caused by too much waste, we need to improve our recycling; and so on. There are any number of plausible arguments about the causes of and the cures for the problem of ecological degradation. However, one can distinguish between more logically consistent and researched propositions and assertions of ideology.

Several ecologists have attempted to develop models of the relationship between human activities and ecological degradation. An example appeared in *Science*, the preeminent scientific journal in the United States, in July of 1997. In this article (Vitousek et al. 1997: 499), several leading U.S. ecologists summarized the ecological impacts of humans on the global environment, developing a model of the direct and indirect effects of humans on the global ecosystem. This model saw the ecological effects of humans as a function of the size of the population, of the use of resources, and of such activities as agriculture, mining, and grazing. Though it is clear that these ecologists understood that somehow human activities were at the base of these changes, it was not clear to them how social processes interacted with the natural environment. At the end of the article, they mentioned the need for further research into the social causes of environmental degradation: ". . . the challenge of understanding a human-dominated planet further requires that the human dimensions of global change—the social, economic, cultural, and other drivers of human actions—be included within our analyses."

It is evident that environmental degradation originates in human activities. Thus, we need to base any model of the causes and cures of this degradation on an understanding of our society and of how our social order developed its current treatment of the natural world.

Social Science and the Natural Environment

One of the first social science models to describe the relationship between social factors and the natural environment was developed by Otis Dudley Duncan (1961: 140–149). Known as the POET model, it stimulated the development of social scientific inquiry into environment and society relations. The POET model sees human societies as being composed of four interrelated components: human *population*, form of social *organization*, natural *environment*, and *technologies* employed by society. A society's impact on the natural environment is seen to be a function of the simultaneous interaction of population levels, social organization, and technological development.

The POET model did not inform any empirical research into the relationships of the various components of the model. This was due to the poorly specified direction of the relationships and to the general nature of the variables, which were difficult if not impossible to operationalize. As a result, this model was underdefined, and testing was not possible (Dietz and Rosa 1994: 279).

The next development regarding the relationship between human activities and environmental degradation was the IPAT model. Developed by Paul Ehrlich and John Holdren (1971), the IPAT model "represents the efforts of population biologists, ecologists, and environmental scientists to formalize the relationship between population, human welfare, and environmental impacts" (Dietz and Rosa 1994: 278). Similar to the POET model, the IPAT model postulates a causal sequence of the environmental *impacts* of human activity on the natural environment, viewing these impacts in terms of three variables: *population, affluence,* and *technological development*. The causal sequence is directional, and the model has shown that it is robust and well defined (and thus an advance from the POET model). There have been a series of well-developed tests and elaboration of this model (Dietz and Rosa 1994, 1997; Rosa 1997). However, the IPAT model falls victim to naturalistic and technological reductionism. Population change and technological development are seen as exogenous to human social organization (Schnaiberg 1980: 61–156; Benton 1994: 28–50). There is also an abundant literature that shows that technology is a social choice, and thus the product of human society (Feenberg 1991; Willoughby 1990).

In essence, the IPAT model views human behavior through the lens of ecology, neglecting the social origins of ecological problems.[1] These limitations have led to calls for the development of models in which the mutual dependence of social practices and ecological conditions are recognized (Dunlap and Catton 1994; Benton and Redclift 1994: 18; Beck 1995: 39).

In the early 1980s a number of new approaches were developed to explain the development of ecological degradation. In *Overshoot* (1980)., the sociologist William Catton argued that the consumption of natural resources by human society had exceeded the carrying capacity of the global ecosystem. The origin of this situation, Catton suggested, lay in the rapid expansion of human population and consumption of natural resources that occurred between 1650 and 1850 (ibid.: 25). In this "Age of Exuberance," the carrying capacity available for human use expanded rapidly as a result of two factors. The first was the discovery of the relatively unpopulated (by European standards) Western Hemisphere, which, when coupled with advanced technology, allowed colonialization and development of these areas by European nations. The second was the development of the knowledge and technology to utilize fossil fuels. Together these factors greatly increased the development and expansion of human society. This extraordinary increase in the availability of natural resources created a cultural belief in the limitless expansion of human society (ibid.: 24). But the two factors on which this expansion was based were one-time events. Hence, as the human population expanded, it caught up with the Earth's carrying capacity. Around 1800 (ibid.: 28; Catton and Dunlap 1980), the total use of natural resources began to exceed the supply. The continued expansion of human society was sustained by drawing down the stored capacity of the Earth's fossil fuels. Thus, the modern social order is based on a temporary and unsustainable level of use of natural resources. However, even as the nonrenewable natural resources on which society depended reached their limits, exuberance continued to be the dominant worldview (Catton 1980: 5). The result of this process is that while the carrying capacity of the Earth's nonrenewable natural resources remained fixed, the development and expansion of human society continued as if the natural resources available for human use were limitless (Catton and Dunlap 1980; Buttel 1986; Merchant 1992).

In a further amplification of this position, the sociologist Riley Dunlap (1992) discussed the effects of environmental degradation on human society. For Dunlap, the natural environment serves three functions for all human societies: as a supply depot, providing the raw natural resources that serve as the sustenance base of the society; as a sink for the waste products of that society; and as the habitat in which the society exists. From this perspective, ecological problems for a society emerge for two reasons. First, the areas used by one function intrude into one another. (An example of this would be the intrusion of the use of the natural environment as a sink for toxic chemicals into an area used for human habitation.) Second, these uses exceed the total capacity that the environment can sustain (ibid.: 711–718). Accordingly, expansion of ecological degradation results in restriction of our living space, depletion of our supply of goods provided by the ecosystem, and overflow of the waste-absorption capacity of the natural environment.

Examining the use of natural resources by human society, Dunlap (ibid.: 717) maintains that, in Catton's terms, we have now entered a post-exuberant age in which the Earth's carrying capacity is being exceeded: "... these functions overlap and therefore conflict with one another much more than in the past, and that their growth may have exceeded the Earth's carrying capacity." Dunlap's analysis is supported by studies in other fields (Ponting 1991; Meadows et al. 1992).

There have been several promising recent steps in the direction of developing a more comprehensive model of the social causes of ecological degradation. In 1992 the National Research Council study *Global Environmental Change* (Stern, Young, and Druckman 1992) provided a major review of the human factors in environmental degradation. The NRC study first establishes a clear distinction between the driving forces and the proximate causes of environmental degradation. The proximate causes are the specific sources of the degradation: urban sprawl, automobile exhaust, greenhouse-gas emissions, and so forth. Most analyses of environmental degradation start with proximate causes, and thus these analyses neglect the social and human causes. To account for human factors, the NRC study defines five variables that are seen as the driving forces of global environmental change: population change, economic growth, technological change, political-economic institutions, and attitudes and beliefs (ibid.: 75).

The weakness of this model is the mixing of ecological concepts, such as population and technology, with distinctly social variables. In addition, many of the social forces that lead to ecological degradation remain unidentified. As Rayner (1992) notes, "the constructionist, or society-driven perspective is entirely ignored." A further development of this model occurred in an article in which Paul Stern maintained that the study of human-environment interactions would need to focus on social origins of environmental degradation, effects of environmental degradation on human society, and feedback between environmental degradation and human actions (Stern 1993: 1897–1899). Dietz and Rosa (1994) have empirically examined these three areas. They have identified the social sources of environmental degradation, and their relation to the IPAT model. Despite their empirical results, they maintain that there is a need to develop models capable of "operationally bridging the differing perspectives between the social sciences and the biological sciences" (ibid.: 278). Dietz and Rosa go on to distinguish between the IPAT model and the social sources of ecological degradation in a society's cultural beliefs, political economy, and social structure. In addition, there is ongoing research into the human responses to ecological degradation, including examinations of the origins of social movements, of increased domestic and international conflict, and of large-scale population migrations (Homer-Dixon et al. 1993: 38; Dobkowski and Wallimann 1993: 198).

Thus, the literature on the origins and effects of ecological degradation encompasses many disciplines and many theoretical perspectives. The three categories of human-environment interactions defined by Stern can be used to define a more parsimonious perspective. These interactions are represented in table 2.1.

The first category of human-environment interactions defined by Stern consists of the human origins of environmental degradation. Two categories of variables are used to analyze these interactions: social variables and human-ecology variables. The social variables that define the origins of a society's ecological problems can be seen to lie in the structure of social institutions, cultural beliefs, and individual behaviors. These social variables can be clearly distinguished from the human-ecology variables, which have been labeled as the driving forces of ecological degradation. The social variables manifest themselves in the society's type of technological practice, its population level, and its levels of resource consumption.

The second category of human-environment interaction defined by Stern focuses on the effects of environmental degradation on human society. The effects on the natural environment take the form of the proximate causes of environmental degradation. These changes affect human society through restriction of our living space, depletion of our supply of goods provided by

Table 2.1
Human-environment interactions.

Origins of environmental degradation		Effects of environmental degradation		Responses to ecological degradation
Social origins	Driving forces	Effects on natural environment—proximate causes	Effects on human society	Feedback through human actions
Social institutions	Population levels	Biodiversity loss	Living space restriction	Government action
Cultural beliefs	Technology practice	Global climate change	Waste repository overflow	Market
Individual personality characteristics	Affluence levels (consumption of natural resources)	Air pollution	Supply depot depletion	Changes
		Water pollution	loss of ecosystem services	Social movements
		Soil/land pollution and degradation	exhaustion of natural resources	Migration
				Conflict

the ecosystem (including loss of ecosystem services), and overflow of the waste-absorption capacity of the natural environment.

Feedback between environmental degradation and human actions forms the third category of human-environment interactions. This category is composed of the human responses to environmental degradation. These can take the form of either intentional activities to modify our activities in response to environmental degradation (e.g., government programs, market changes, or social movements) or unanticipated responses (e.g., migrations or conflicts). These activities then impact in a recursive manner on the social origins of environmental degradation.

This scheme can display the different approaches and interrelationships between social and ecological variables by providing a general model of the relevant processes. As such, it is a decontextualized model that does not situate the process of ecological degradation within a specific social, cultural, or historical dynamic. Accordingly, this model fails to elaborate on the aspects of the current social order that contribute to the process of ecological degradation. This limits the utility and relevance of this approach. As Beck (1995: 39) has noted, "this debate is either senseless or vacuous, and probably both, unless one also considers society's power structures and distributions structures, its bureaucracies and its prevailing norms and thought patterns." To expand our understanding of the process of ecological degradation, it is necessary to delve further into the social processes that influence social institutions, cultural beliefs, and individual personality structures. Then, on the basis of this understanding, we can begin to chart a scheme from which we can understand the social origins of ecological degradation.

To develop this framework requires a working knowledge of the processes of social order and social change. One of the most comprehensive perspectives in the social sciences, and one that may be able to unify this work, is the approach known as *critical theory*.

The Social Order and the Theory of Communicative Action

Jürgen Habermas's work, which spans more than 40 years, has been guided by a desire to realize the Enlightenment's goal of the emancipation of human individuals and society through the creation of a rational and just society. As was noted above, the aim of critical theory is to facilitate understanding of our society and to enable us to use this knowledge to construct a rational and moral society. This project is not just of academic interest. Rather, it is born of a deep sense of the inadequacy of our current social order to solve the problems we face, including ecological degradation.

Habermas has attempted to reinvigorate social thought and to connect it to our current situation. His work can be seen as an attempt to provide Western society with the cultural means to reorganize itself. His social theory develops a theory of justice from which competing claims about what actions we should follow can be decided. To develop this theory of justice, he develops a communicative ethics based on the links between rationality, democracy, and justice. Another role of social theory is to attempt to create a coherent worldview to serve as the basis for public deliberation about what actions should be taken. To meet this task, Habermas has developed a complex model of social interaction and of how the continued expansion of the market and the state administrative bureaucracy progressively threatens the communicative structuring of the everyday world. This process also expands into the natural world, bringing on more ecological degradation.

By examining Habermas's model, we will be better able to form an expanded model of the relationship between social dynamics and ecological degradation.

Habermas's scheme progresses from the fundamental presuppositions of speech to the formation of coherent worldviews, or discourses. These coherent worldviews serve as the basis for social organization, which form the components of a social order. Thus, Habermas's work builds from the everyday use of language to the formation of discourses and social organizations. These discourses and social organizations become institutionalized and form an interrelated system that constitutes a society. These institutions then evolve and adapt to changing conditions through communicative action. Not only does Habermas's framework explain the development and change of social order; it also provides a framework in which the rationality and morality of different social orders can be compared.

Language and Communicative Action

The theoretical framework Habermas uses is the philosophy of language. Following the linguistic turn in social theory (Simon 1990), the philosophy of language focuses on the creation of joint social action in everyday life through communication and language use.[2] Analysis of the social order from a linguistic perspective starts with the viewpoint of the participant in the everyday lifeworld, in which we coordinate our actions by talking with one another (Habermas 1987a: 126). Through the use of language in the lifeworld, we develop a notion of reality (Brown 1983: 135).[3] The establishment of a shared reality creates a definition of the situation on which joint action is based. Habermas (1996: 323) examines how this process of communication creates and maintains social order by first developing a model or thought experiment of social interaction consisting only of "pure communicative sociation." He calls this model the *ideal speech situation.* In this model, only mutual agreement based in communication forms the basis for joint action. This type of interaction defines communicative action, which is "interaction in which all participants harmonize their individual plans of action with one another and thus pursue their aims without reservation" (Habermas 1984: 294). In this situation, it is assumed that the actors can express their goals truthfully and without reservation, that all pertinent evidence can be brought into play in the discussion, and that the agreement is based on reasoned argument (Habermas 1983a: 4).

This model of the ideal speech situation allows Habermas to identify the "functionally necessary resources" (1996: 325) for communication to exist. He maintains that these presuppositions enable the communication process and are embedded in the use of any natural language. He starts by noting that there are three dimensions of human existence. We exist as beings in a physical world, as social beings within a cultural world of shared meanings, and as individuals with unique identities. To engage in join action, we must coordinate our actions in each of these dimensions (Habermas 1983a: 5).[4]

We use language to create a mutual agreement on which action can be based (Habermas 1998a: 247). If we are to come to an agreement about what action should be followed, a persuasive case must be made that the statement can provide valid answers to challenges in three dimensions: First, to establish a correspondence with the human experiences in the physical and natural world, the statement must accurately represent the objective world—i.e., it must be true. A statement is proved to be true by providing empirical referents to establish its credibility. Second, to integrate the proposed action into the cultural beliefs, the statement must make an argument that the acceptance of a proposed action is in accordance with other existing cultural norms and beliefs—i.e., that it is a moral and legitimate action.

This is accomplished by engaging in moral argumentation. Finally, to establish a link between the statement and the individual's subjective (inner) world, the statement must be an authentic representation of the speaker—i.e., the speaker must be truthful. This is determined by establishing whether the individual can be trusted to carry out the action as stated. This validity claim can only be met by comparison of stated intention with actual behavior over time. Hence, the development of an intersubjective agreement regarding the nature of the specific reality must fulfill three validity claims: truth, normative rightness, and sincerity. By fulfilling these three validity claims, an intersubjectively accepted statement is constructed which describes the relevant states or events in a situation, places the interaction within the moral criteria of a specific historical and social context, and ties the individual's personal identity to the interaction (Habermas 1984: 136).[5] By agreement of two or more persons, the basis for rational social action is formed. Thus, language enables joint social agreement through fulfilling the validity claims of normative rightness and truth through practical and theoretical discourse, and truthfulness through congruency between words and actions. This definition of the situation thus forms the basis for social action (Habermas 1996: 4).

Discourse and Social Organization

Our conversations about the world accumulate into specialized languages that deal with specific areas. These cumulative languages form a stable definition of the situation, which is known as a discourse. A discourse is "a formation constituted by all that is said, written or thought in a determinate field" (Teymur 1982: 2). It provides the basis for a common interpretation of the situation. Contained in each discourse is a commonly held stock of practical social know-how. By using this pre-given image of social reality, the members have a basis for acting together in an organized manner (Brown 1978: 373–374; Brown 1990; Brulle 1994). This includes rules and paradigmatic examples of how to enact different social roles (Dietz and Burns 1992; Burns and Dietz 1992).

   The formation of a discourse defines a range of human interaction that falls within a particular definition of the situation. According to Emirbayer and Goodwin (1994: 1438), this regularized and limited pattern of behavior constitutes a social organization: "Organizations result from the development and instantiation of a discourse as a legitimate reality in a bounded network of action. The establishment of a legitimate discourse is embedded in a bounded network of communicative action in a co-generative process." In satisfying the validity requirements for rational intersubjective agreement, communicative action creates and maintains symbolic structures. These structures enable the construction of social order by providing the symbolic resources used in forming and maintaining groups (Habermas 1984: 308). Thus, a discourse provides the cultural basis from which stable behavioral expectations originate, joint action is undertaken, and organizations are formed.[6]

A discourse constitutes a cultural structure with its own "internal logic and organization" (Emirbayer and Goodwin 1994: 1438). Every discourse has a narrative structure.[7] By explaining the past and pre-structuring future action, a discourse creates a basis for coordinated human action. The development and establishment of a legitimate and binding discourse is an achievement of human communication that constructs social reality and constitutes both individuals and organizations (Brinton 1985: 281). Thus, the construction of a discourse and a social organization is a communicative act of human agency and a practical accomplishment of a human community.

Systems of Social Order

Habermas argues that social organization must be viewed as both a lifeworld and a system. As previously noted, the everyday lifeworld is social order from the viewpoint of the participant in everyday life. Linguistic communication in the lifeworld creates and socializes individuals into a joint social reality, and thus performs the task of creating and maintaining social organizations. However, by itself this theoretical perspective lacks a notion of how our actions create and maintain large-scale social formations, such as the market or the state (Cohen 1982). To maintain a stable social system, the boundaries of social organizations must be defined and maintained. In addition, the relationships between different institutions must be integrated. Finally, the entire social order must develop external stability in relation to its environment (Habermas 1991: 252).

The function of these social processes is outside of everyday understandings of action in the lifeworld. To view these phenomena, Habermas (1987a) maintains, one must adopt a perspective outside the lifeworld. To shift to view society as a whole from the outside, systems theory forms a useful perspective to make these social processes visible.[8] Thus, Habermas's examination of social order proceeds from a two-tier model of the social order, the tiers being lifeworld and system.

A systems viewpoint sees society as a series of related institutions which act together to ensure the integration and stability of society as a whole. The systems aspects of social order are derived from the evolution of the lifeworld. Through a historical evolutionary process, a means for stabilizing and integrating the results of human actions has developed. Habermas describes this development of social structures apart from the everyday understanding as both a historical creation and a part of the evolutionary logic of social learning.[9] Social systems arise from the development of new worldviews and practices through communicative action in the lifeworld. The process of testing validity claims establishes a pragmatic testing of social practices and outcomes. This increases the stock of knowledge in the society and represents an expanded cognitive potential available for social use. These practices can then be regularized and institutionalized (Habermas 1998a: 250). If these practices are subsequently institutionalized, they replace communicative action with institutionalized criteria, or special codes, which Habermas (1991: 258) labels "steering media." This reduces the burden on communicative action, and creates stable organizational routines (ibid.), thereby making possible the creation of subsystems of society in the form of complex functional networks. As the complexity of the social order increases, these processes of testing validity claims become separated into distinct disciplines. As a result, the knowledge particular to the various spheres develops as separate and specialized disciplines that retain the society's knowledge in distinct realms. Empirical knowledge about the world becomes the discipline of science. Science ensures the provision of valid knowledge on which to base action. The definition of legitimate social relationships becomes the province of the discipline of law, and the discipline of self-representation becomes institutionalized in the discourse of aesthetics, which enables the development of authentic adult personalities. Each of the specific disciplines provides the knowledge resources necessary to sustain social order. The development of these discourses then proceeds through debates over the validity claims in each area.

These systems of society which are derived from our lifeworld conversations retain their correspondence to the knowledge claims on which everyday language is based (Habermas 1998a: 247). The relationship between everyday language and the systems of social order is summarized in table 2.2.

The first row of table 2.2 deals with the creation and maintenance of the symbolic structure of *culture*. For a human community to act together, a system of valid and mutually shared knowledge must be sustained. Corresponding to the validity claim of truth, the symbolic system of culture ensures mutual understanding. Communicative action serves to transmit and renew cultural knowledge. By securing a "continuity of tradition and coherence of knowledge" (Habermas 1987a: 140), the task of cultural reproduction is performed. Valid and rational knowledge adequate for mutual understanding is provided. Communicative action also ensures the reproduction of knowledge needed for social legitimization and for the socialization of adult personalities.

The second symbolic structure created by communicative action in the lifeworld is *society*. A human community is based upon the establishment and sustaining of interpersonal relations in the form of legitimate social orders. This symbolic system enables social integration and the establishment of solidarity by "coordinating actions by way of legitimately regulated

Table 2.2
Language and symbolic components of social order (adapted from Habermas 1987: 142).

Validity claim of speech	Outcome of communicative action	Social task performed	Symbolic component of social order
Truth	Valid knowledge	Cultural reproduction by ensuring rationality of knowledge	Culture
Normative adequacy	Legitimate relationships	Social integration by maintaining solidarity of members of society	Society
Sincerity	Responsible personality	Socialization of a responsible adult personality	Person

interpersonal relations and stabilizing the identity of groups" (Habermas 1987a: 140). It creates a situation of legitimately ordered interpersonal relationships, defines a series of mutual moral obligations, and creates legitimate individual social roles. This symbolic system enables social integration and the establishment of solidarity.

Finally, the symbolic structure of *the person* enables the formation of personal identities. Corresponding to the validity claim of sincerity in speech, this symbolic structure links the individual to social order. Communicative action serves the formation of personal identities by "securing the acquisition of generalized competencies for action and seeing to it that individual life histories are in harmony with collective forms of life" (Habermas 1987a: 141). The socialization of society's members is performed in the lifeworld through the development of the personality system of the individual. Communicative action develops the interactive capabilities of the person, the motivations for individual actions according to socially defined norms, and provision of the cultural knowledge for self-expression. This forms a responsible adult personality that is capable of a rational reconstruction of self and society through agency (ibid.: 98).[10] Thus, it is through communicative action that the "symbolic structures of the lifeworld are reproduced by way of the continuation of valid knowledge, stabilization of group solidarity, and socialization of responsible actors" (ibid.: 137).

These three social institutions develop and adapt to changing conditions through the use of communicative action (Habermas 1979, 1990). The process of redeeming validity claims allows social institutions to change in response to new circumstances. This allows for the implementation of increased learning potential as social arrangements are subjected to validity testing. From Habermas's viewpoint (1987a: 145), a rational society is one in which the social institutions "come under conditions of rationally motivated mutual understanding, that is, of consensus formation that rests in the end on the authority of the better argument." Thus, social order evolves from and is changed by communicative action in the lifeworld.

Rationality, Communicative Ethics, and Constitutional Democracy

The process of developing mutual agreement in the everyday lifeworld defines a moral framework known as *communicative ethics*. As was noted above, the rational formation of joint social action occurs through communications that fulfill the validity claims of truth, morality, and sincerity, thus forming a basis for rational action by establishing a mutual consensus. Rationality is thus tied to intersubjective communication carried on in the conduct of life.[11]

The fulfillment of validity claims also defines a communicative ethics. In conversations in a human community, the claims of the speaker must be validated for the discourse to be rational. The process of validation requires an open speech community in which the unforced force of the better argument prevails. This ethical relationship is a presupposition of mutual communication. Implicit in this relationship of mutual communication is a basic recognition and acceptance of others. This includes a respect for the autonomy and integrity of the other person's identity and selfhood (Schlosberg 1998: 68). Based on this argument, Habermas (1996: 458) has posited that proposed actions be evaluated under the following discourse principle: "The only regulations and ways of acting that can claim legitimacy are those to which all who are possibly affected could assent as participants in rational discourses."

As the Western social order developed, the complexity and the extent of behavioral expectations increased. The development of the modern constitutional democracy was a social innovation to reduce the complexity of interactions while incorporating the legitimacy of communicatively achieved consensus. Habermas builds from this insight to develop a theory of deliberative politics and law that examines the "normative elements which are always tacitly presupposed when we participate in these democratic and constitutional practices" (Carleheden and Gabriels 1996: 2). Owing to the complexity and degree of interactions between strangers in our society, the costs of developing mutual understandings through communicative action make it impractical. In addition, while moral norms might provide guidance on how to act in a certain manner, they provide only weak motives toward their realization. However, law provides a means to regulate these interactions (Habermas 1998b: xii). Collectively agreed upon behavioral expectations become institutionalized as collectively binding laws. The legal code contains the legitimate and reciprocally binding behavioral expectations within which social institutions operate.

The legitimacy of binding laws is due to their origination as the outcome of a discursive process through which citizens engage in self-determination through open and rational discourse. This requires that citizens always be able to understand themselves as authors of the laws to which they are subject. Thus, communicative ethics takes the shape of a "self organizing legal community" (Habermas 1996: 326). This ties the rationality of an open discourse to the formation of legitimate laws in a democracy (ibid.: 297).

The development of constitutional democracy can be seen as an attempt to implement communicative ethics in a political system (ibid.: 327). This requires that individual citizens have the basic individual rights of freedom of speech, individual legal protection, and freedom of association. These constitutional rights create a space in which citizens can act as autonomous agents. In addition, to fully realize the ethical requirements defined by communicative ethics, there is a need for political rights. Political rights are necessary to ensure broad and inclusive public participation in the making of collectively binding laws that regulate social interaction.

To guarantee the provision of political rights, Habermas (1996: 458) develops his Democracy Principle, which maintains the legitimacy of legally binding behavioral expectations in a constitutional democracy. This principle states that the establishment of the legal code, which is undertaken with the help of the universal right to equal individual liberties, must be compelled through communicative and participatory rights that guarantee equal opportunities for the public use of communicative liberties. This Democracy Principle thus forms a guide for "reconstructing the network of discourses that . . . provides the matrix from which democratic authority emerges" (ibid.: 5). This provides a norm from which the rationality and legitimacy of political decision-making institutions can be judged. Habermas argues that "the sociological translation of the discourse theory of democracy implies that binding decisions, to be legitimate, must be steered by communication flows that start at the periphery and pass through the entrance to the parliamentary complex or the courts" (ibid.: 356).

A rational and legitimate social order thus requires development of both individual and political rights. In order to realize communicative ethics, the political constitution must "institutionalize those communicative presuppositions and procedures of a political opinion and will-formation in which the discourse principle is applied" (ibid.: 458). This requires political institutions that take the form of free association of citizens, the unrestricted flow of information and opinion into the press, and collectively binding decisions made and enforced by representative government institutions (ibid.: 118–131, 305–306).

The theory of communicative action is based on the idea of the ideal speech situation. However, this form of pure communicative action is only a model; it does not fully deal with the nature of actually existing communication of "finite, embodied actors" (ibid.: 324). While it can identify the necessary conditions that enable communication, it does not consider the contingently imposed constraints on this process (ibid.: 323–328). Thus, Habermas maintains, "no complex society could ever correspond to the model of purely communicative social relations" (ibid.: 326). However, communicative action sustains the vitality of social institutions of both the self and organizations through discursive redemption of the claims of the propositions of each discipline. This results in providing adequate knowledge to ensure the continuation of valid knowledge, the stabilization of group solidarity, and the socialization of responsible actors. Accordingly, the rationality of a social order can be evaluated on the basis of "the standards of the rationality of knowledge, the solidarity of members, and the responsibility of the adult personality" (Habermas 1987a: 141).

Historical Development of the Western Social Order

From his theoretical and normative framework, Habermas conducts an examination of the Western social order. For Habermas, the cumulative advancement of the project of modernity over 500 years has resulted in a series of complex social relationships that define this social order. This process can be viewed as the differentiation of the lifeworld into specific institutions that guide various forms of action. Strategic actions that aim to bring about a given end in the world were institutionalized in the market and the administrative state. Normative actions that aim to realize a certain ethic were institutionalized in representative constitutional government and positive law. Aesthetic actions that seek to present authentic expressions of the self were institutionalized in art.

However, this development was not uniform. The market expanded and became the dominant social institution, over both legal constitutional democracy and aesthetic institutions. As the modern social order developed, the implementation of new procedures of material reproduction required division of labor. Coordination of production and exchange via traditional or communicative action was replaced by a nonlinguistic means of coordination: the market. Similarly, administrative power developed as a means of stabilizing the effects of the economic system. This allowed the economic and administrative institutions to decouple themselves from the everyday lifeworld (Habermas 1987a: 186–187; 1991: 250–261). This replaced linguistically arrived at behavioral expectations with ones derived through the steering mechanisms of money and power (1987a: 183). As the market and the state expanded, the parallel development of normative and aesthetic discourse into similar institutions failed to take place. Modernization is thus a process of one-sided and contradictory rationalization. Moral and aesthetic reason form subjugated discourses that have little effect on the development of action in the dominant social institutions. Thus, the economy and the state, directed by the nonlinguistic media of money and power, are the dominant components of modern society.

The development of these two means of coordinating action was legitimized through two separate logics in liberal capitalism. The market took on the appearance of a natural process, legitimized by the "justice inherent in the exchange of equivalents" (Habermas 1973: 22). The state simplified the amount of moral reasoning an individual had to engage in as society became more complex, and it substituted positive law for interaction to regulate interactions between strangers (Habermas 1996: 452). As was noted above, these laws were legitimized through appeal to universal values developed through democratic will formation. This separation of legitimizing schemes between the market and the state was made possible because economic class relations were no longer directly instituted through political power as they were in feudal society. Rather, they took the "anonymous form of the law of value" (Habermas 1971: 68) in the form of market outcomes. This depoliticized relations of domination and control by the state (Habermas 1973: 22 1991: 256).

The Project to Socially Restrain Capitalism

The market and the state did not remain separated. Since the Great Depression, the state has become an active participant in economic activities in a project that can be seen as the social restraint of capitalism. The state has increasingly intervened in the economy to stabilize the economic system and to compensate for the adverse effects of capital accumulation. This has resulted in the current situation, in which the state is an active participant in economic activities and also assumes responsibility for ensuring the legitimacy of the market. Economic growth is dependent on expansion of both investment and consumption. In the current mixed economy, to maintain economic growth, the state takes actions to encourage investments and to expand consumption (Offe 1984; Offe and Ronge 1982; O'Connor 1973: 24).

The state's effort to stabilize the market economy is based in the notion of the positive state. Using empirical social science to identify the levers that control the mechanisms of society, the state develops and implements programs of social engineering (Brulle 1994; Etzioni 1976; Goodwin 1975; Camhis 1979: 56–72). The aim of this project is to ensure the smooth functioning of the market economy while contributing to the overall improvement of society.

Although these actions have apparently stabilized the market and minimized the worst aspects of the business cycle, they have also undermined other aspects of the social order. This dynamic increases the difficulty of maintaining the social order. At the core of this process is the repoliticization of the market. As the state increasingly intervenes in the economy, the separation of the market and the state is destroyed. The intervention of the state in the economic process to avoid economic crisis and to ensure capital accumulation results in the politicization of market activities (O'Connor 1973: 13–39). The legitimacy of the social order is instead based on the outcomes of both the market and state action (Habermas 1971: 68). This makes the state responsible for maintaining social stability and ameliorating problems that create sufficient political opposition to the existing order.

The action of the state also undermines the logic of private ownership. The instruments used by the state to encourage investment also serve to limit the control of the owners of capital in the investment process. The market is increasingly dependent upon politics for creating and maintaining the capital-accumulation process (Offe and Ronge 1984: 254–256). This inability of the market to maintain itself creates a politically maintained private market in which socialized production and private appropriation of production exist in a system legitimized by formal democratic rules. This creates a conflict between the normative justifications for collective decisions. Market outcomes are legitimized as the outcomes of democratic will formation. This leads to a series of contradictions and crises in Western society.

On the one hand, the state depends on economic growth to provide it with the funds necessary to accomplish its tasks. The legitimacy of the government rests on maintaining economic prosperity and the state's flow of funds. This causes the state to adopt policies that support capital accumulation. On the other hand, state power is legitimized and maintained through the formal rules of political democracy. Owing to the democratic heritage of the Western state, and the successful struggles to gain universal political rights in modern Western nations, state action can be legitimized only through formal democracy. However, the legitimacy of private control over the production process in the marketplace is based on the now-obsolete belief in a self-legitimizing market. Thus, the state is required to maintain a supposedly free market through political action. To do this it must represent the interest of capital accumulation as being in the general interest.

The active intervention of the state also undermines the motivational beliefs of the individual. The active participation of an individual to maximize his or her life chances through the market is premised on the acceptance of market outcomes as legitimate. For this motivational system to operate, personal success or failure in the market economy must be seen as the result of individual effort. However, because economic outcomes are seen as a result of both individual effort and political decisions, the tendency to accept market outcomes as fully legitimate is decreased (Offe and Ronge 1984: 256). The questioning of the validity of the link between an individual's efforts and market outcomes undermines the motivation to unquestioningly participate in the market. Instead, a general attitude of cynicism develops, characterized by opportunistic behavior, and the result is a general withdrawal of self-identification with the market process (Offe and Ronge 1984; Habermas 1971, 1984, 1987a).

The repoliticization of the market creates a situation in which officials of the state are required to act politically to maintain the conditions of the "private" market and to maintain the legitimacy of social and market outcomes. This requires the state to take action to ensure the growth of the market and, at the same time, meet the effective political demands of the underemployed, the unemployed, and other state clients and compensate for the social and environmental costs of economic growth. However, state institutions lack the political autonomy, the resources, and the organizational capabilities necessary to address these demands (Frankel 1982: 263). In general, the results of state action can be characterized as awkward and changing interventions that reproduce the existing patterns of social relations. As shown by more than two decades of evaluation research on state policy, the "expected value of the effect of any program hovers around zero" (Rossi et al. 1979: 341). Hence, the state acts only as a crisis manager, engaged primarily in deflecting political criticism and not engaged in solving fundamental problems.

To reduce the extent of politically effective demands requires the insulation of state action from public input, for in acting to realize and legitimize capital accumulation the state must act to realize particular interests, and the legitimacy of such actions in a process of open communication would be questioned. If these actions were decided upon in a situation of open communication, this "would bring to consciousness the contradictions between administratively socialized production and the continued private appropriation and use of surplus value" (Habermas 1973: 36–37).

The actions of the state are insulated from public scrutiny by four processes.

First, access to institutional policies is limited through the use of a scientific discourse in the development and analysis of government policy. By creating a technocratic, value-neutral discourse, it removes moral considerations from public policy and limits public input. "The scientization of politics," writes Habermas (1970: 68), "reduces the process of democratic decision-making to a regulated acclamation procedure for elites alternately appointed to exercise power." Without the power to speak the specialized discourse of science, the lifeworld is without a voice, and citizens are reduced to the status of a population to be managed by the state.

The second process is the management of political demand through the adoption of neocorporatist decision making (Outhwaite and Bottomore 1993: 114; Schmitter and Lehmbruch 1979: 13). In this process, the state involves selected groups in negotiations to enhance its legitimacy and to mobilize populations to support certain lines of action. Included in this closed process are representatives of interest groups that have the potential to disrupt the functioning of social institutions. The price of being allowed to engage in such negotiations is the adoption of "reasonable" positions and predictable and orderly behavior by group members. This results in the creation of limited demands as the outcome from these types of negotiations, thus creating a form of top-down manipulation of interest groups (Schmitter and Lehmbruch 1979; Manes 1990: 60; Szasz 1992: 524).

The third process that insulates state agencies from public scrutiny is their organizational structure. The requirement to facilitate capital accumulation is incorporated into the normal organizational routine through establishing the agencies' objectives and range of legal authority, controlling the size and content of its budget, and establishing lines of authority and accountability (Zeitlin 1980).

Finally, insulation of the state from public demands is accomplished through increasing the decision-making power of the executive agencies of the state. Through the passage of a largely symbolic and administratively discretionary bill, the political demand is transferred from the visible public arena to the less-visible bureaucratic agency (Hayes 1978: 149), so that established interest groups can use their political power with less fear of effective political opposition. As a result, "political questions become defined as technical or administrative problems requiring experts rather than politicians for their solution, and, not incidentally, insulating the interests involved from popular pressures." The "only opposition to it appears to be political and illegitimate intrusion of politics into supposedly social neutral state agencies." (Zeitlin 1980: 27–28)

Hence, there is very little public about public policy. The cumulative result is atrophy of the public sphere. The public sphere was historically envisioned as an arena in which the common good was debated and a democratic consensus was reached. This process would result in "the communicative generation of legitimate power" (Habermas 1992a: 452) for a social order. Instead, the public sphere is reduced to "the manipulative deployment of media power to procure mass loyalty, consumer demand, and compliance with systemic imperatives" (ibid.). The result of this process is the removal of substantive decision making from direct access by the public while retaining the form of democratic institutions. Thus, "the citizenry, in the midst of an objectively political society, enjoy the status of passive citizens with only the right to withhold acclamation" (Habermas 1973: 36).

While certain aspects of the project to socially restrain capitalism have succeeded, it has essentially created a crisis-ridden social order. This has resulted in a situation in which a number of economic, political, cultural, environmental, and individual motivational problems all occur simultaneously (Offe and Ronge 1984: 254–256).

Commoditization and Juridification of the Lifeworld

In addition to creating a crisis-ridden social order, the modern social order restricts its ability to renew itself by restricting its basis in communicative action. As a result, the institutions of the social order are unable to renew themselves. This results in deterioration of the lifeworld.

The everyday lifeworld is threatened by the one-sided and distorted development of rationality in the modern social order. Instrumental reason forms the dominant rhetoric of legitimization of both market and state actions.[12] The atrophy of the public sphere seals off the lifeworld from providing input into the administrative state and the market. This creates a social order dominated by the narrow instrumental rationality that defines the imperatives of the state and the market. Fulfilling the instrumental imperatives of these two institutions forms the dominant legitimization scheme for social action, unrestrained by moral or aesthetic rationality.

Without any effective controls, the market and the state expand into the everyday lifeworld in the manner of "colonial masters coming into a tribal society—and forcing a process of assimilation upon it" (Habermas 1987a: 355). The substitution of economic relations for what were previously lifeworld practices (child rearing in day care centers, fast food substituting for meals at home, and so on) shows the increasing commodification of our daily lives.

The parallel expansion of administrative procedures designed to structure civil society in accordance with the state's requirements (for example, social work, old-age pensions, sexual and moral education in the schools) shows the continued expansion of the state into areas that were previously under the control of the lifeworld. Habermas (1987a: 356–372) characterizes this process of state expansion as "the juridification of the lifeworld." More and more aspects of social relations become subject to administrative control. As the process of juridification expands, the state expands and the ability of the lifeworld to create and maintain meaning becomes diminished (ibid.: 330).

The negative effect of the penetration of lifeworld destroys its capabilities to generate meaning. As the complexity and the influence of the institutions of the market and the state increase, the ability of a member to understand and influence them decreases. The processes by which these systems operate is not a theme that can be developed from linguistic processes and interpersonal relations. Hence, the ability of the members to understand and control their situation is restricted. Instead, the imperatives of the market and the state guide everyday life. This creates a systematic distortion of communication in the lifeworld "in such a way that the interrelation of the objective, social and subjective worlds gets prejudged for participants in a typical fashion" (Habermas 1987a: 187).

Since "meaning can neither be bought nor coerced" (Habermas 1991: 259), the market and the state cannot renew the social institutions. This results in a series of crises in culture, in politics, and in the sustainability of our lives in the lifeworld (Habermas 1987a: 386). No longer is conflict in society primarily about what types of actions will be taken by the state or market. Instead, it is primarily about whether the basis for action will be the imperatives of the market and the administrative state or whether it will be lifeworld communication (Habermas 1985: 342).

A society that undermines its capacity to form and reform its basis in linguistic communication cannot sustain itself. If the ability of the lifeworld to reform social order is blocked, a number of social functions are not adequately performed, leading to the creation of social pathologies. The relationships between the symbolic structures that create and maintain social order, the social processes performed by each structure, and the resultant social pathologies are illustrated in table 2.3.

The task of cultural reproduction requires the creation and maintenance of the symbolic structure of culture. A system of rational (valid and mutually shared) knowledge must be sustained for a human community to act together. The legitimacy and adequacy of this knowledge is ensured through communicative action in the lifeworld. However, the commodification and juridification of the lifeworld blocks the testing of social systems. As was noted above, the market and the state are contradictory and are based on two opposed logics of action. Without testing, this irrational system of legitimization is not challenged. However, gaps in its adequacy appear. These gaps are not corrected, and the legitimacy of this worldview is not sustained. A coherent and valid worldview is not transmitted throughout society. This

Table 2.3
Social processes and pathologies (adapted from Habermas 1987: 143).

Social process	Culture	Society	Person
Cultural reproduction by ensuring rationality of knowledge	Loss of meaning	Withdrawal of legitimization	Crisis in orientation and education
Social integration by maintaining solidarity of members of society	Unsettling collective identity	Anomie	Alienation
Socialization of a responsible adult personality	Rupture of tradition	Withdrawal of motivation	Psychopathologies

leads to a loss of meaning, or to the creation of an incoherent or unworkable worldview. This is also experienced in society at the normative level by a general withdrawal of belief in the social order. For the person, this creates problems in the education and socialization of an integrated adult personality.

Social integration requires the establishment and sustenance of interpersonal relations in the form of legitimate norms of behavior. Lifeworld communication establishes a mutually shared set of norms on which behavior can be based. This creates a situation of legitimately ordered interpersonal relationships, defines a series of mutual moral obligations, and defines legitimate individual social memberships. This symbolic system enables social integration and the establishment of solidarity. The one-sided dominance of the market and the state effectively hinders consideration of the normative consequences of action on an equal basis with economic and political considerations. The atrophy of the capabilities of the lifeworld to generate a moral notion of the common good has resulted in a situation in which the narrow instrumental rationality of state and market imperatives forms the dominant legitimization scheme for social action. As a consequence, the current social order is composed of a world of narrow, competing interest groups, without a common image of the "good" and consequently without an overall vision of social direction. This results in a social order without an adequate normative basis, and it creates confusion regarding what the "whole" or collective identity stands for. In the society, a lack of any notion of an adequate collective moral order results in normlessness, social relativism and anomie. The individual is left apart from society and alienated from any notion of a larger moral order.

The socialization of society's members is performed in the lifeworld through the development of the personality system of the individual. A self that is in accord with the social order is constructed through three processes: learning the culture in which the self participates, understanding and believing in the normative basis of this community, and becoming capable of expressing oneself well enough to participate meaningfully in this community. The outcome is a socially responsible adult personality capable of functioning in this social order. As a result of the failure of the social order to provide either valid cultural knowledge or a normative basis for the creation of social solidarity, the socialization of the self into an adult personality is restricted. Without adults that are adequately knowledgeable about the social order's cultural knowledge, the tradition is not passed along to the succeeding generations; the result is a rupture in the tradition. Additionally, without a normative consensus the personality has no guide on which to base its pattern of motivation. The outcome of this is an extremely restricted personality that is an active construction of the market (Adorno and Horkheimer 1944; Ewen 1976; Barnouw 1975).

The modern personality develops around two values that are crucial to fitting the self into the market: consumerism and possessive individualism (Habermas 1987a: 325). Consumerism creates a personality that is easily manipulated by advertisements promising that the good life will be found in the next commodity. This leads to an unending quest for consumption in an anxious search for social status (Veblen 1899; Simmel 1907; Weber 1904). Possessive individualism manifests itself in narcissism, leading to a life divorced from civic involvement and to a single-minded careerism that exalts the satisfaction of personal ambition as the preeminent value. The individual is left apart from society and alienated from any notion of a larger moral order. With the atrophy of the lifeworld, the development of human interactions based upon a shared moral vision becomes more difficult to form and sustain. This means the gradual decay of relations of mutual trust, reciprocity, and a genuine identification with the other. Unable to synthesize the current situation, consciousness reverts to outmoded traditions that no longer can make claims as valid knowledge or are hopelessly fragmented into a series of unrelated local language games (Bellah et al. 1985). Habermas (1987a: 355) concludes that "in place of false consciousness[,] we today have a fragmented consciousness that blocks enlightenment by the mechanism of reification."

This analysis shows how the isolation of the lifeworld restricts social learning. This results in the crisis-ridden nature of current social order. The failure to fulfill these processes of the construction of social order results in a lack of rational knowledge on which the dominant reality is based, in decay of the solidarity of members of the society, and in atrophy of the development of an adult personality capable of acting as a member of the community.

Habermas's analysis of the development of Western society provides a basis for understanding the processes that are threatening the lifeworld and disrupting the development and renewal of social order. Based on this analysis, Habermas argues that a reinvigoration of democracy is essential to enable us to correct our social institutions and to deal with our numerous social and political problems, including the continued process of ecological degradation.

As this review of Habermas's work shows, his intellectual project makes two substantial contributions. First, his communicative ethics provides a theory of justice from which competing claims about what actions we should follow can be decided. Based on this ethics, Habermas seeks to provide a standard for the creation of a moral and rational society. He also provides a comprehensive model from which a coherent worldview of current Western society can be constructed. This can then serve as the basis for public deliberation about what actions should be taken. Together, these contributions provide intellectual resources that can be used to reorganize our social order to better deal with its current circumstances.

The "Green" Critique of Critical Theory

As I have noted, there has been an extensive debate regarding the theoretical framework developed by Habermas. Within environmental social theory, the most rigorous debate about the applicability of Habermas's framework to "green" politics centers on the work of Robyn Eckersley.

Eckersley accepts much of the content of communicative ethics. She agrees that community decisions should be reached through democratic deliberations among representative of the various groups of the society. She also agrees with Habermas that the participants in this dialogue should seek to build social solidarity with the other members of their community by striving to develop a mutual understanding of one another's position (Eckersley 1999: 27).

According to Eckersley, communicative ethics, being founded in human speech acts, cannot integrate concern for nonhuman nature. Critical theory considers only the development of norms between mutual participants in a discourse, thus omitting consideration of the moral concerns of other species that are not capable of participating in this dialogue. Nature is not a subject of morality in communicative ethics. Thus, as currently constituted, critical theory is an anthropocentric belief system that separates and privileges human emancipation over the emancipation of nonhuman beings. This limits the application of this perspective to informing a cultural practice that could create a sustainable society that preserves and protects biodiversity. Eckersley argues that we need to develop an expanded ethics so as to include the nonhuman community in our decision-making process. Based on this concern, she seeks to extend communicative ethics to mandate inclusion of the concerns of nonhuman nature in our decision-making processes.

First, Eckersley argues that moral concern is not limited to individuals who have the ability to participate in community dialogue. For example, the concerns of infants, who cannot speak, are routinely represented in human discussions by their representatives. Eckersley argues that ability to engage in the conversation is not a requirement to be treated as a subject of moral concern. All the beings affected by a given action have a right to be treated as moral subjects, regardless of their level of communicative competence. Thus, Eckersley argues (1999: 46) for the following addition to Habermas's discourse principle: "A just common structure of political action must be common to all those affected, irrespective of whether they happen to be able to speak or gesture."

Second, Eckersley argues that the various living entities that make up nature have the capacity to define themselves through evolutionary processes, and thus have their own integrity as centers of agency (ibid.: 42). Thus, they are self-defining beings and have a right to be treated as moral subjects. Accordingly, nonhuman beings should participate in our deliberations about proposed actions in the form of this question: "If they could talk and reason, would they agree to the proposed norm?" (ibid.: 44)

However, there is a problem in designating a particular human group or individual to serve as a representative of nature. Since all our knowledge of nature is socially constructed and particular, there is no true or authentic representative of nature within the human community. "The upshot is that no one can speak for 'nature in itself'; we can only speak about the nature we humans have constituted." (ibid.: 40) Eckersley's proposal to meet the moral requirement of including the concerns of nonhuman beings in human decision making is to mandate institutional procedures that would "guide human decision makers away from putting 'the silent environmental constituency' at grave risk" (ibid.: 45–46). She then turns to the Precautionary Principle[13] as one such principle that would institutionalize the moral mandate to consider the impacts of human actions on nonhuman beings.

Eckersley develops a strong and elegant argument for including the concerns of nature within a politics guided by communicative ethics. However, her argument would be more compelling if it addressed the following concerns.

First, it is not at all clear that an "ecocentric" perspective would result in the creation of the type of society desired by Eckersley. The anthropological literature contains several examples of serious ecological disruptions created by nomadic peoples who viewed nature as sacred (Ponting 1992: 18–36). How the desired type of decision-making system would be institutionalized and realized in our current social order, and what impacts it would have, can only be projected with a high degree of uncertainty. Thus, it is by no means certain that this is a realistic and practical strategy.

Second, the attribution of agency, and hence the status of moral subject to nature is problematic. Agency is not a term commonly used by ecologists to describe the behavior of different species. Instead, action is attributed to instincts, genetic predispositions, and for some higher mammals, learned behaviors. In addition, Eckersley does not specify the level at which the agency of nature takes place. There are many different definitions that could be used here, such as ecosystem, species, subspecies, or individual animals. What aspect of nature has agency, and whether there is one "interest of nature" for the multiple species that exist in nature are questions not addressed by Eckersley. Eckersley also ignores the socially constructed nature of her own arguments. Hazelrigg (1995) argues that there is no universal position from which to experience or know nature. Human society develops its knowledge of nature through carrying out practices in the world. In doing so, we experience nature in different ways. Included in our experience are the recalcitrant and repeating experiences that form the reality of the natural world. To label these experiences "the agency of nature" is to endow them with a particular characteristic that is socially constructed. The natural world emerges from the encounter of human society with the "limits of its agency and autonomy" (Hazelrigg 1995: 160). However, these "realities" emerge only in response to particular practices. Since all experiences of nature are particular and historical, only particular knowledge of aspects of nature can emerge. Thus, Hazelrigg (ibid.: 12) convincingly argues that our definition of nature is "tied to our existence as a community that is engaged in carrying out practices in the world." Thus, the construction of the idea of nature is not a neutral activity; rather, it is a rhetorical act based in particular interests and perspectives. An argument for the existence of agency in nature is not a universal and objective idea, as Eckersley would have it. Rather, it is a particular social construction: "Preserving nature means preserving a particular construction of what nature is supposed to be." (ibid.: 292) Eckersley constructs a nature endowed with agency and then seeks to derive a moral warrant for an ecocentric norm. In doing so, she essentializes her argument and falls victim to ethical naturalism (Caterino 1993).

Third, Eckersley misrepresents the rigorous foundations of communicative ethics and tries to develop a parallel argument for the acceptance of an ecocentric ethics. She argues that the basis for the acceptance of communicative ethics is that this ethics provides the best means for ensuring a free dialogue that protects the autonomy and integrity of human actors. She then argues that an ecocentric ethics should also be accepted because it would work best to protect nature. This argument ignores the robust foundations of communicative ethics in Habermas's analysis of speech acts. One of his major intellectual projects was to develop a non-arbitrary link between democracy and reason. Habermas has expended considerable effort in on analyzing speech acts, and he has developed a reconstructive science of the presuppositions of speech. While communicative ethics clearly defines normative requirements, the basis for its acceptance is the empirical analysis of speech acts that fills the entire first volume of *The Theory of Communicative Action* (Habermas 1984). By ignoring that foundational work, Eckersley reduces the cognitive force of the argument for acceptance of a communicative ethics and produces a weak and arbitrary argument for the acceptance of an ecocentric ethics.

In addition, one can question the need for the development of an ecocentric ethic at all. Several scholars, including Habermas, have argued that communicative ethics already includes Eckersley's concern with the mandate for an ecological norm. Communicative ethics focuses on how a human community can arrive at rational, legitimate, binding norms. Nowhere does it specify what these norms should be. Habermas has suggested a manner in which this type of moral argument would be extended to encompass human interaction with the natural world (1993: 111): "In that case animals would come to be recognized in all situations as possible participants in interaction, and the protection to which we feel obligated in our interactions with animals would be extended to include protecting their existence." This does not mean that nature attains the status of a moral subject. However, valid arguments for the ethical treatment of nature can be made using aesthetic, moral, and practical reasoning. What this implies for arguments valuing nature is that there are different forms of reasoning about the value of the natural environment, and we need to make each argument clearly. We need all three forms of argument to develop a rational (i.e., competent, moral, authentic) discourse that can be the basis for the creation of a social order capable of sustaining the biodiversity of the natural world. Hence there is no theoretical barrier to the development of an ethics for the treatment of nature through the process of human dialogue (Torgerson 1999: 118). Using an expanded notion of accountability, Caterino (1994) has attempted such an extension of communicative ethics to the natural world. Caterino argues that the good life is something that we decide upon and carry out as part of enacting our self-identities and selves. He argues that this norm is founded on recognition of the interdependence of humanity and nature. Because of this mutuality, a norm for human behavior can be developed that includes ecological considerations in the definition of the good (ibid.: 32). On the basis of this extension, communicative ethics can be seen as encompassing norms that include the self, society, and nature. Torgerson (1999) expands this argument, arguing that in an undistorted communication situation the irrationality of the belief of the ethics of human domination of nature would be recognized and delegitimized. A rational discourse focusing on the ethics of nature would then be part of the discourse of a society guided by communicative ethics: "Human beings participating in a rational form of life could not reasonably avoid considering the moral dimensions of their relationship with nonhuman nature. To do so would be to focus on the difference between the human and the nonhuman, ignoring the significance of the continuity."[14] (ibid.: 120)

Finally, there is no guarantee that Eckersley's ecological ethic, if adopted and institutionalized in law, would result in the protection of nonhuman nature. This has already been tried in the United States. The National Environmental Policy Act of 1969 (42 USC 4321) mandated the government to preserve and protect the natural environment by considering the impacts of all its actions on nature. This was followed by the Endangered Species Act in 1973 (16 USC 1531), the purpose of which is to "provide a means whereby the ecosystems upon which endangered species and threatened species depend may be conserved" (16 USC 1531(b)). While these laws have made some difference in the preservation of nonhuman species, biodiversity loss continues to accelerate.

Thus, it is not enough to develop and institutionalize an ecological ethic in law. There is no absolute guarantee that this norm would be accepted and enforced in the human community. Rather, this is a matter of human agency and action. To enable the natural world to be protected, we need to hold our social institutions accountable for their actions. This means that, as Dobson (1993: 198) states, summarizing Habermas's position, "healing the rift between human beings and the natural world . . . is not a matter of joining what was once put asunder, but of getting the relations between human beings right first."

There is no necessary conflict between an ecocentric norm and communicative ethics. Rather, there is a relationship of mutual reliance. Developing and legitimizing an ecocentric perspective requires the opening of the public space, and it requires democratic dialogue to promulgate, gain acceptance of, and enforce such a norm. Enactment of a communicative ethics is thus a necessary prerequisite to the creation of an ecologically sustainable society (Torgerson 1999: 123). Thus, the project to extend communicative ethics to include the voices of nonhuman beings is misplaced, and the proposed alternatives are unworkable. The legitimate concerns of "green" theorists for the ethical treatment of the natural world can be integrated into the perspective of critical theory.

Conclusion

Questions about preservation of the natural environment are not just technical questions; they are also about what defines the good and moral life, and about the essence and the meaning of our existence. Hence, these are not just academic or technical matters, to be settled in elite dialogues between experts. These are fundamental questions of defining what our human community is and how it should exist. In a democratic society, this requires the participation of all the citizens in the discussion, which entails shifts in the conditions under which decisions about the natural environment are made to ensure that all voices, including those that speak for nonlinguistic beings,[15] are heard. In this way, we can develop the means by which our society can adjudicate these competing value systems in a rational and democratic manner and can develop a dialogue about biodiversity that is competent, moral, and authentic.

Habermas's work provides a theory of justice from which competing claims about what actions we should follow can be decided, including how the natural world should be treated. Thus, the theoretical perspective developed by Habermas creates a theory of justice and rationality that can serve as the normative basis of a sustainable society. In addition, this perspective forms a coherent worldview that can provide a basis for an expanded model of the relationship between social dynamics and ecological degradation. Habermas's analysis of our current situation provides a comprehensive model from which an understanding of the causes and the cure of the process of environmental degradation can be approached. Together, these contributions provide valuable intellectual resources that can be used to inform a politics that can contribute to the creation of an ecologically sustainable society. This model can be used to develop an understanding of the causes and means to address ecological degradation.

3

The Social Dynamics of Environmental Degradation

Environmental problems are not problems of our surroundings, but—in their origins and through their consequences—are thoroughly social problems, problems of people, their history, their living conditions, their relation to the world and reality, their social, cultural and political situation.

—Ulrich Beck (1986: 81)

Our society's attempts to deal with environmental degradation have been characterized as "little more than cosmetic" (Pointing 1991: 400). Our social and political institutions have been unable to solve our ecological problems. Gradually we have become used to accepting a degraded natural environment. Photochemical smog, polluted water, urban sprawl, and toxic-waste dumps have become regular parts of our everyday experience. We can, of course, escape to remnants of a clean and livable environment in our parks and wilderness areas. However, this only highlights our consciousness of the extent of the damage we have done to the communities in which we live.

If one accepts the basic outlines of Habermas's analysis of our current situation developed by Habermas, several questions remain. How does this model apply to environmental degradation? Can this perspective assist in the development of a plausible explanation for the genesis of environmental degradation? Can it provide some guidance as to how degradation can be reversed? For the most part, the social origins of ecological degradation have not been addressed within the framework of critical theory. There is, in the field of environmental sociology, an extensive body of literature on the social creation of environmental problems. Yet, while the environmental sociology literature points to the social origins of ecological degradation, there is little unanimity in this field about which social processes are important. The theory of communicative action may be able to provide a position from which these various approaches can be synthesized. In this chapter, I attempt to sketch a model able to combine the multiple perspectives of environmental sociology on the causes of environmental degradation into a more comprehensive approach through the use of the theory of communicative action.

Environmental Degradation and Social Processes

There are a number of sociological perspectives that locate the origins of ecological degradation in the characteristics of our social order. This has been noted by the environmental sociologist Riley Dunlap (1992: 708): ". . . environmental problems are inherently social problems, for they are created by society, are having an increasingly negative impact on society, and will require coordinated social action to be solved." Combining the POET and IPAT models, Dunlap develops a model of the three specifically social variables that are theorized to drive ecological degradation: social institutions, cultural beliefs, and individual personality characteristics.

Dunlap's model parallels Habermas's analysis of the study of social order. The theory of communicative action views social order as comprising three forms of action. A particular form of social theory studies each type of action. By focusing on the process by which speech ties together the objective, social, and subjective worlds, the theory of communicative action allows for examination of the different types of social action and of the concomitant school of sociological thought (Habermas 1984: 84–141). Different forms of social thought also appear in environmental sociology. Thus, the theory of communicative action can form a perspective from which the various approaches to the social origins of the problem of ecological degradation can be viewed.

The first type of action identified by the theory of communicative action is *teleological*. In this type of action, human actions aim to bring about some end or goal. The social world is seen as a network of interrelated social institutions. The application of this perspective in environmental sociology thus sees the origin of environmental degradation in our social institutions.

The second type of action is *normatively regulated*. In this type of action, the actor seeks to orient his or her action to common values. The social world is seen as an intersubjective cultural system of mutual beliefs and norms in which the actor is embedded. Environmental sociology based on this approach sees our cultural beliefs as the origin of ecological degradation.

The third type of action is *dramaturgical.* In this type of action, the actor selectively presents a version of his or her self to an audience. The social world is seen to be a theatre in which the actor plays different roles. The study of ecological degradation from this viewpoint focuses on the creation and change of the dominant personality structures and behaviors in society.

Each of these approaches models a particular aspect of social interaction. The theory of communicative action provides a basis for the simultaneous production of social order through the coordination of action thorough social institutions, the adherence to shared cultural beliefs, and the socialization and maintenance of stable personality characteristics. However, Habermas traces the pathologies of our current social order to the expansion of the market and the state into the lifeworld.[1] The isolation of the lifeworld restricts social learning and results in the crisis-ridden nature of current social order. Habermas's analysis can be extended to develop an understanding of the social processes that create ecological degradation.

The continued expansion of the state and the market into the lifeworld is the core social process that destroys the natural environment. Working within their own internal logics, these institutions result in expropriation of the natural world without regard for its ecological or biological consequences. Just as the colonialization of the lifeworld progressively destroys the ability of our society to reproduce its symbolic and institutional components, the same processes also threaten to destroy the society's physical and biological basis. These processes are summarized in table 3.1, which sketches an extension of Habermas's analysis of the dynamics of our social order to examine the social processes that create and maintain the process of ecological degradation. From this viewpoint, there are three distinct structures that contribute to ecological degradation.

Social Institutions and Environmental Problems

The social sources of environmental problems are defined by Dunlap (1993: 725) to include "all aspects of social organization." The social organizations that are seen as major causes of environmental problems in this type of analysis—the market and the state—are contradictory and are based on

Table 3.1
Social sources of environmental problems.

Social structures	Cause of ecological degradation	Outcome
Social institutions	Irrational and contradictory discourses form dominant basis for action	Ecological impacts not considered in development of state and market action
Cultural beliefs	Market and state internal imperatives replace normative considerations	No ecological ethics bounds human activities
Individual personality characteristics	Careerism and possessive individualism form dominant personality style	High levels of consumption and lack of capacity to engage in joint action

two opposed logics of action. This creates a situation in which a series of contradictory actions are taken by the state to ensure the continued functioning of the market. This creates ecological degradation through the actions of both economic and political institutions in the form of market and state failure.

A number of studies have examined the origin of environmental problems in the political economy of advanced capitalist economies (Schnaiberg 1980; Schnaiberg and Gould 1994; Cotgrove and Duff 1980; Cotgrove 1982; O'Connor 1973, 1984, 1987). Based primarily on neo-Marxist economic analysis, these approaches assign primary blame for the origin of environmental problems to the logic of the market. Alan Schnaiberg, an environmental sociologist who has conducted extensive analyses of the relationship between the capitalist economy and the environment,[2] argues that the capitalist economy forms a "treadmill of production" that continues to create ecological problems through a self-reinforcing mechanism of ever more production and consumption. The logic of the treadmill of production is an ever-growing need for capital investment in order to generate goods for sale in the market. From the environment, it requires growing inputs of energy and material. When resources are constrained, the treadmill of capitalist production searches for alternative sources rather than conserving and restructuring production. The treadmill operates in this way to maintain a positive rate of return on investments. In theory, the state is responsible for reconciling disparities between the treadmill and the needs within the society. In practice it has often acted to accelerate the treadmill in the hope of avoiding political conflict (Schnaiberg 1980: 418). The ecological result of this process is that the use of natural resources continues to increase, regardless of the consequences for the sustainability of the ecosystem.

This analysis is extended in the work of Carlo Jaeger. For Jaeger (1994: 115–129), the creation of the market replaced action based upon mutual accountability established in conversations with strategic interaction based on the exchange of goods. As the market expanded into more areas, it transformed the means of meeting human needs into the ends of human existence. This process resulted in the separation of means from ends and in the establishment of distinct normative fields unrelated to conversational accountability (ibid.: 123). It is this process of the progressive destruction of human accountability that forms the basis of the destruction of the environment. Jaeger (ibid.: 113) argues that "when human agency degenerated into labor, the relationship between humankind and the biosphere, too, became marked by irresponsibility." This has created an economic system that operates without regard to its external environmental effects (ibid.: 188). The basis for this systematic disregard of the environmental consequences of economic activity is the social rule that establishes a positive rate of return on investment (ibid.: 200).

Another analysis of the relationship between social structure and the development of the problem of ecological degradation[3] is that of Ulrich Beck (1986), who provided a model of the interaction of technology, social dynamics, and ecological degradation. For Beck, the continued development of industrial production is based on the dynamic of modernization and industrialization. Beck characterizes these processes as "blind and deaf to consequences and dangers" (ibid.: 28).

At the center of modernization is the application of scientific research and knowledge to expand economic growth. The power to define technological development, and thus our future, becomes concentrated in the private corporate power that controls and directs much research and development. This results in a shift in the locus of power from the nation state to the corporations and their control over the scientific agenda. Beck (1995: 73) notes that "when everyone awaits the contours of another society from the application of microelectronics, reactor and genetic technologies . . . it no longer makes sense to locate politics only in the political system." Driven by the need to maximize profits, corporations continue to develop new technologies that produce unforeseen risks for the entire society. As a result, industrial risks change quantitatively and qualitatively. They are no longer limited in time or space, aftercare or restoration is not possible, monetary compensation is not possible, and the causality or blame for the accident cannot be defined. Thus, accountability for the accident cannot be established, and these industrial projects are not insurable (Beck 1996: 31; Beck 1995: 22–23). This breaks down the ability of society to ensure the safety of its citizens from the production of industrial hazards. Thus, the security of citizens against harm from industrial processes is no longer maintained (Beck 1995: 22–23). This generalization of risks that are not limited in space or time creates a phenomenon that Beck calls the "End of the Other." In the course of human history, one group of people inflicted violence on the "other" (an enemy, a scapegoat, a dissident). Now, the harm caused by global environmental problems, such as global climate disruption or ozone depletion, is inflicted on all persons, regardless of social class or ethnicity (Beck 1995: 27). In addition, there arise winners and losers in the politics of the distribution of environmental harm (Beck 1986: 53). This leads to the unequal distribution of environmental degradation to relatively powerless communities. "What is denied," Beck writes (1995: 29), "collects itself into geographical areas, into 'loser regions' which have to pay with their economic existence for the damage and its unaccountability." This creates a "risk society" in which the politics of eliminating scarcity to provide equity and freedom from want is overlaid by a politics based on minimizing anxiety and eliminating risk by providing safety (Beck 1986: 49). The creation of a "risk society" through continued economic growth and the application of science leads to questions arising about the previously legitimizing ideas of industrial society and progress, including just means for compensating for industrial risks, the notion of scientific truth, and the identification of economic development with progress.

The impartiality and rationality of science is challenged in a risk society (Beck 1986: 29). The supposed ability of science to develop objective knowledge is demystified by the confrontation of the problems that its application has brought about. Science is seen as a social construction that not only solves problems but also creates them. Its legitimacy as an impartial and objective guide to action is delegitimized by the "failure of techo-scientific rationality in the face of growing risks and threats from civilization" (ibid.: 59). So the practices and assumptions of the scientific enterprise becomes the subject of critique (ibid.: 155–174; Harre et al. 1999: 11).

In addition, the previously unquestioned linkage between economic development and progress is challenged. Economic growth creates wealth for some and imposes risks on others. The problem of ecological destruction returns to its creators in a boomerang effect. The risks of modernization catch up with those who create them. Property becomes devalued through ecological destruction (Beck 1995: 60). The politics of a risk society challenges the fundamental premises on which industrial society is constructed. "In those conflicts," writes Beck (ibid.: 40), "what is at stake is the issue of whether we can continue the exploitation of nature and thus, whether our concepts of 'progress" 'prosperity', economic growth,' or 'scientific rationality' are still correct. In this sense, the conflicts that erupt here take on the character of doctrinal struggles within civilization over the proper road for modernity."

According to Martin Janicke, not only does the market fail to take into account the ecological consequences of its actions; the state also fails to control the market. Janicke develops the concept of *state failure* to explain the inability of our government to address the problem of ecological degradation. Janicke (1990: x) maintains that the evidence for state failure is "the inability of governmental reform policies to replace the outmoded postwar pattern of industrialism," which, he maintains, lies at the basis of the problem of ecological degradation. State failure is due to the tight relationships that develop between the government bureaucracies and industries, and to the relative exclusion of the public institutions that are supposed to hold the bureaucracies accountable (ibid.: 14–30). As a result, the response to ecological degradation takes the form of post facto responses rather than anticipatory and preventive actions (ibid.: 41–54).

The imperatives of the market and state are the dominant logics guiding our social institutions action toward the natural environment. Together, Schnaiberg, Jaeger, Beck, and Janicke illustrate how, despite some minor and generally ineffective legal requirements to include this information into decision making, our dominant social institutions—the

market and the state—operate without regard to the ecological consequences of their actions. This disregard is one of the key factors in the
destruction of the natural environment. Joel Whitebook (1979: 61) argues
that "the ecological crisis is, at its roots, caused by the strain that the
incessant expansion of the economy—which enlists science and technology for its purposes—places on the natural environment." The fact that
the lifeworld has no effective control over these processes limits the rationality of the social order. Together, these four authors present a powerful
case for the social creation of environmental problems. At the core of their
analyses is the idea that the dynamics of the market and the state are systematically unable to take into account the effect of capital accumulation
on the environment.

Cultural Beliefs and Environmental Problems

Dunlap (1993: 724) identifies cultural beliefs as a second social source of
environmental problems. In our society, the cause of ecological degradation is seen by many social theorists to be caused by deeply held cultural
beliefs.[4] Meeker (1972: 6) describes the underlying assumptions of this type
of analysis:

The origins of environmental crisis lie deep in human cultural traditions at levels
of human mentality which have remained virtually unchanged for several thousand years. The premises upon which our culture has been built are powerful and
durable, and their weight upon us must be appreciated before we can hope to alter
their structure.

A wide range of cultural factors, including the Western worldview, patriarchy, and Christianity, are considered to be sources of ecological degradation.[5] Perhaps the most widely used notion of this cultural belief in
environmental sociology is defined by the term *Dominant Social Paradigm*
(Milbrath 1984, 1989; Dunlap and Van Liere 1978; Olsen et al. 1992),
which defines a collection of beliefs regarding nature, planning, economic
growth, social change, and politics that constitute the core orientations in
modern society toward nature. The Dominant Social Paradigm sees man
as having hegemony over nature and the natural environment as a limitless
resource available for exploitation (Catton and Dunlap 1980).

Roderick French (1980: 45) characterizes Western philosophy as exem-
plifying "human chauvinism." Our dominant philosophical traditions, he

suggests, see humanity as separate from and discontinuous with the rest of natural life. He maintains that this illusion should be replaced with a philosophy that connects us with the natural environment:

> . . . the shock of the ecological critique of the humanities may have been precisely what was needed in order to force us to face up to the philosophical bankruptcy of our traditional assumptions as they have played themselves out in the twentieth century. The absurdity of a vision of human existence thoroughly detached from man's natural context suggests that the recovery of a sense of purpose might well come through a rediscovery of our connectedness with the natural environment.

In a recent analysis of the cultural beliefs that we use to develop our ethics toward the nature environment, the philosopher Rom Harre argues that "there is a misfit between the linguistic resources and the problems to be addressed" (Harre et al. 1999: 50). Harre goes on to argue that the Western scientific discourse contains a number of blind spots, and that it is an insufficient means to understand and form the basis of our relationship with the natural environment (ibid.: 159).

While these analyses yield valuable insights into the cultural origins of ecological degradation, they do not explain why we cannot develop an ethic that would protect the natural environment. To explain this situation, the theory of communicative action can provide a useful framework. If the normative basis of our social order were based in lifeworld communication, we could develop new ethical norms which would guide our actions toward the natural environment. The research of Harre and other cultural analysts would assist us in expanding the range of ideas considered and would provide an increased stock of cultural resources for our use in developing an ecological ethic.

However, the one-sided dominance of the market and the state effectively hinders consideration of the normative consequences of action on an equal basis with economic and political considerations. The atrophy of the capabilities of the lifeworld to generate a moral notion of the common good, including an ecological ethic, has resulted in a narrow instrumental rationality based on the imperatives of the state and market as forming the dominant legitimization for social action. Instrumental reason effectively blocks the creation of a normative agreement regarding the preservation of the natural environment. Instead, the pursuit of means replaces the development of moral and ethical ends of action, blocking the formation of a normative consensus on which ecologically rational actions can be based.

Personality Characteristics and Environmental Problems

The third social origin of environmental problems identified by Dunlap (1992: 726) is individual personality characteristics. The personality characteristics that are seen to be the primary cause of ecological degradation are possessive individualism and careerism. Habermas argues that, owing to the failure of the social order to provide either valid cultural knowledge or a normative basis for the creation of social solidarity, the socialization of the self into an adult personality is restricted. Additionally, without a normative consensus, the personality has no guide on which to base its pattern of motivation. An extremely restricted personality that is the active construction of the market is the outcome. This takes the form of the dominant personality styles' being based on careerism and possessive individualism.

Beck (1986: 90) argues that continued modernization dissolves traditional forms of life through their "institutionalization and standardization." Specific historical developments, including shifts in the labor market, job mobility, competition, and educational standardization, have created a situation in which "people have lost their traditional support networks and have had to rely on themselves and their own individual fate with all its attendant risks, opportunities and contradictions" (ibid.: 92). As a result, "the individual becomes the reproduction unit of the social in the lifeworld" (ibid.: 90). This existence takes the form of a standardized collective existence of "isolated mass hermits" (ibid.: 32). Social isolation leads to the definition of the self becoming a project of the individual in the form of a career, and the manipulation of the self through advertising (Beck 1995: 40, 59).

These personality types contribute to environmental problems in three areas. First, the stress on possessive individualism fuels consumerism in the infinite quest to acquire the material goods that constitute a socially appropriate "lifestyle" as developed by Madison Avenue. This never-ending consumption leads to an ever-increasing use of resources to support this type of personality. Second, the penetration of the market and the state into the lifeworld fragments consciousness and reifies social structures. This blocks recognition of the social causes of environmental problems. Instead, an individualist notion of the cause of environmental problems develops. This redefines the cause of environmental destruction to the level of individual choices. It develops actions (such as recycling) to address this problem, inhibiting the development of social action at the group level. Finally, the self-definition project of the "career" manifests itself as the "bureaucratic personality" (Riesman 1950) or the "organization man" (Whyte 1956). This personality type creates a dominant behavior based on fitting into, rather than changing, social institutions. This blocks the formation of a responsible adult personality that could act as an agent to recreate new social alternatives in conjunction with others. Taken together, these personality characteristics encourage high levels of material consumption and limit the capacity of individuals to engage in joint action.

Social Resolution of Environmental Problems

To address the destruction of our biological and physical environment, an increase in social learning capability is needed. Since the creation of environmental problems originates in our social order, their resolution lies in the development and implementation of alternatives. Specifically, this involves replacing the limited and one-sided rationalization of society based on instrumental reason to a full rationalization of strategic, normative, and aesthetic action. A social order is adaptive to the extent that the tasks of cultural reproduction, social integration, and socialization are met.[6] Hence, the reconstruction of an ecologically sustainable society would be based on integrating ecological concerns into all three symbolic structures of the social order. This process is shown in table 3.2. This table shows the need for social change that would perform three tasks: institutionalize ecological rationality, establish an ecological ethic, and develop individual capacities to participate in a democratic and sustainable society.

Institutionalize Ecological Rationality
The existing decision-making process limits input from the public and allows the internal imperatives of the market and the state to form the dominant basis for instrumental action. This distortion of communicative action results in the failure the decision-making process in both the state and the market to take ecological information into account. Consequently, the existing decision-making process fails to adequately address environmental problems. Thus, to increase the rationality of knowledge in institutional decision making, a

Table 3.2
Social resolution of environmental problems.

Social structure	Cause of ecological degradation	Outcome	Corrective action
Social institutions	Irrational and contradictory discourses form dominant basis for action	Ecological impacts not considered in development of state and market action	Institutionalize ecological rationality
Cultural beliefs	Market and state internal imperatives replace normative considerations	No ecological ethics bounds human activities	Establish and legislate ecological ethics
Individual personality characteristics	Careerism and possessive individualism are dominant personality characteristics	High levels of consumption and lack of capacity to engage in joint action	Develop personality characteristics of "ecological citizen"

means must be found to ensure that ecological information is adequately considered in decision making in both state and market institutions. The inclusion of ecological concerns in the collective decision-making process is based on the reform of existing social institutions along more democratic lines through a process of institutional experimentation (Brulle 1994: 115–117; Fischer 1992, 1993). Calling for the creation of a more democratic society, Serge Taylor (1984: 3) argues that the process of integrating ecological information into our institutions is "inextricably a question of how we fashion our political and administrative institutions."

The importance of public participation in developing rational and just decisions that include concern about the natural environment has been argued by numerous authors.[7] Examining the role of public citizens in hazardous-waste decision making, Morgan (1993: 29) concludes that public participation is essential for ensuring the "general welfare." The link between public participation and rationality is recognized in the National Research Council's study on risk decision making (NRC 1996: 23–24): "... relevant wisdom is not limited to scientific specialists and public officials and that participation by diverse groups and individuals will provide essential information and insights about a risk situation."

One of the more cogent analyses of societal decision making regarding the physical and biological environment is offered by John Dryzek in his 1987 book *Rational Ecology*. After a review of the efficacy of several social choice mechanisms for the resolution of environmental problems, Dryzek concludes that a situation of open communication offers the best possibility for countering "ecologically irrational mechanisms" (245–247). In a similar manner, Ulrich Beck argues for the democratization of techno-logical development and economic structures. For Beck (1986: 228–235), the key to addressing the problem of ecological degradation is to remove the causes of risk production. The first component of Beck's strategy is the democratization of techno-economic development. This would involve attributing accountability to those who create risks through the political control over corporate research-and-development decisions. Before a technology could be developed, its ecological harmlessness would have to be demonstrated publicly (ibid.: 63). The second component of this project would be a revision of scientific practice. Instead of being steered by corporate funding and the need for research to provide further economic development, science would "develop independent, theoretically sound alternative perspectives that demonstrate and illuminate the sources of problems and their elimination in industrial development itself" (ibid.: 176). This would require scientific autonomy (ibid.: 82).

Opening up the decision-making process to democratic participation would create a social situation that would allow for testing of the validity of the information on which collectively binding decisions are made. Just as scientific knowledge progresses through vigorous debate and the generation and testing of alternative theories, a more open social order would allow for more ecological knowledge to be incorporated into decision making. This would increase the use of ecological information and the overall rationality of the social order.[8]

Establish and Legislate Ecological Ethics

The establishment of a public participation is a necessary, but not sufficient action to reconstruct an ecologically sustainable social order. Effective solutions to our most pressing environmental problems, such as global warming and biodiversity loss, require co-operative human efforts on a global scale (Apel 1980: 228). Yet no effective ecological ethics currently bound our activities. Thus, our cultural traditions need to be transformed to serve as effective guides for developing these norms (Brulle and Dietz 1992).

As table 3.2 shows, the one-sided rationalization of our social order has resulted in the atrophy of moral considerations into strategic actions, where means are transformed into ends. To replace this narrow logic of legitimization, ecological norms capable of serving as a moral bound on the activities of the state and the market must be developed and then made binding by being encoded into law and effectively enforced (Buttel 1986: 350).

An effective ecological ethic must develop the capacity to motivate worldwide social action. It must create a means of allowing the vast diversity of cultural beliefs to work toward a common goal (Cooper 1996: 257). To accomplish this, an ecological ethic must include knowledge of physical, biological, and social sciences. In addition, an ecological ethic must acknowledge the wide range of cultural viewpoints, including conflicting notions of what is sacred and profane, what constitutes truth and heresy, and even basic notions regarding what it means to be human (Falk 1986: 77).

No one cultural tradition will be able to accomplish this task. Rather, to meet this requirement will require the development of multiple forms of ecological ethics to motivate action in different cultural systems. The theory of communicative action specifies that to take joint action, the subjects involved must agree with one another that the proposed action is moral. There is no requirement that joint action be based on one set of cultural beliefs, only that there exist good normative reasons for acting in a particular manner. As Schlosberg (1998: 87–101) notes: "Political unity does not require that a political agreement be reached based on identical reasons. Rather, unity can be achieved through recognition and inclusion of multiplicity and particularity."

The specific ethical arguments used to address this need have grown into an expansive literature that cannot be adequately summarized here.[9] The theory of communicative action cannot provide guidance regarding which of these different ecological ethics should be adopted; that is a matter to be decided by the participants themselves. However, it can contribute to this debate by defining the social conditions in which a morally binding ethic can be constructed and enacted (Habermas 1983: 1–20).[10] To create this ecological ethic, we need to create a number of public spheres in different cultural systems. These public spheres must be outside of the domination of the control of the market and state to allow for the development and promulgation of various ecological ethics.

Develop Personality Characteristics of an "Ecological Citizen"
The third row in table 3.2 illustrates the self-construction process. A socially adequate personality system is based on developing the knowledge, the moral motivations, and the interaction capabilities to ensure that the self is in harmony with the social order (Habermas 1987a: 142–148). To function as a member of a democratic and ecologically sustainable society would involve the development of individual personality characteristics that would replace possessive individualism and careerism with the notion of an "ecological citizen." Building on the model of moral development and socialization developed by Habermas, Kastenholz and Erdman (1992) define the capacities an ecological citizen would have to acquire. Creation of this type of personality system would require the acquisition of sufficient ecological knowledge to participate meaningfully in discussions of collectively binding actions, the internalization of ecological ethics that legitimize the creation of an ecologically sustainable society, and the development of personal capabilities to participate in this form of social order.

To create this type of personality structure requires self-development and learning at the individual level in two areas. The first area is cognitive enhancement. In the process of individual learning, the understanding of the world changes from a world centered on the immediate physical presence of the body to a world understood in terms of abstract symbols and reversibility of positions within discourse. This process of expansion of worldview involves the acquisition of both technical knowledge and knowledge of the cultural values of the other members of society. Cognitive enhancement would provide the individual with the knowledge necessary to engage in competent and meaningful discussions with the other members of the society about what actions should be undertaken. The second area of individual learning is through moral development. This defines the process by which individuals "come to be able to make judgments about right and wrong" (Webler, Kastenholz, and Renn 1994: 2). Moral development follows a logic from egocentric norms based in a narrow calculation of self-interest to universal norms based in a community of dialogue

(Habermas 1979, 1990). To engage in a mutual and democratic dialogue requires sufficient moral learning to enable one to engage is the discursive validation of particular courses of action. This requires the development of a responsible adult personality in which "the actor adopts toward his own actions, the evaluative perspectives of the expert, the generalized other and his own self all at once" (Claus Offe, in Habermas 1992a: 454).

To bring about the personality characteristics of an "ecological citizen" requires that individuals acquire the necessary education in both technical and cultural areas. In addition, this knowledge must be integrated into the actions of the individual. This requires the individual to participate in collective decision making. In their analysis of the effects on the individual of participation in public participation programs in Switzerland, Webler, Kastenholz, and Renn (1994) argue that, in developing the capacities to participate in this dialogue, a process of cognitive enhancement of the individual occurs. In addition, moral development is enhanced through public participation. Hence, the development of individual capacities to participate in a democratic and ecologically sustainable society can be enhanced through public participation.

Thus, the construction of a social order that is in harmony with the natural environment requires actions that cut across all the existing aspects of society. There is no single set of actions or a program that can resolve this problem. What is required is a fundamental restructuring of the current social order to institutionalize ecological rationality, develop and legislate an ecological ethic, and create a personality structure of the ecological citizen.

Civil Society, the Public Sphere, and Representative Organizations

The above analysis demonstrates the social origins of ecological degradation and thus the need for social change to reverse the process. Since the market and state, through their internal dynamics, have shown a very limited self-correction capability in regard to their ecological impacts, an increase in the capability for social learning must be developed outside these institutions.

As the examination of the means to resolve the problem of ecological degradation shows, democracy is a key component in enabling this process of social change. First, the solutions to ecological problems are not just technical in nature. There is no universal position from which policies can be developed. Input is needed from all citizens to inform a comprehensive picture of the situation (Stern 1991). Second, norms are legitimized and accepted through debate and participation. To develop a legitimate ecological ethic requires the active and full participation of the citizens. Finally, the creation of a new personality structure requires socialization via participation in new worldviews and belief systems.

This links the resolution of ecological problems to the development of a democratic social order.[11] Since social learning is based on open communicative action in the lifeworld, the enhancement of the democratic will-formation capacity of society is a necessary prerequisite for initiation of the actions through which environmental problems could be resolved. The question then becomes: How can this democratization be accomplished? As Habermas (1991) notes, "we are faced with the problem of how capabilities for self-organization can be developed." Both the market and the state are based on a limited notion of instrumental reason, and thus unable to effectively generate adequate creative alternatives to satisfactorily address the problem of ecological degradation. There is a need for a plausible and practical theory of transition from our current state to this desired state. Beck (1995: 34, 75–76) sees social change originating from multiple agents struggling at a number of locations within and between institutions and the lifeworld, and in the development of the capacities of institutional self-criticism by fostering the development of whistleblowers and debates within institutions between experts. Yet how this is to be accomplished is hazy and ill defined.

After reviewing the work of several leading social theorists who have examined the environmental problematic, Goldblatt (1996: 202) concluded that "these social theorists and indeed the contemporary ecology movements, have yet to find the public idiom and institutional format in which that argument, in the face of structural resistance and public apathy, becomes utterly compelling." Accordingly, a political theory of transition must be developed that can generate alternatives and enable a politics to satisfactorily address the problem of ecological degradation.

Civil Society
One such theory of transition looks to institutions outside the dynamics of the market and the state as sites where these actions would originate. Offe

(1990: 76) notes that associative relations based in informal organizations can serve as a "beachhead" for the development of the lifeworld's direction of the market and the state. Habermas and a number of other social theorists have pointed to the institutions of civil society and the public sphere as sites in which democratic social change could originate.[12]

Civil society is composed of voluntary institutions that exist outside of the direct control of the market and the state (Habermas 1992a: 454).[13] The network of organizations includes specialized organizations, including "clearly defined group interests," "associations and cultural establishments," and "public interest groups and churches or charitable organizations" (Habermas 1996: 355). Because these organizations are based in communicative action, the institutional dynamics defined by market forces and the political power of the state are minimized in their operation. This independence forms the key to the capacity of civil society to serve as a site for the generation of democratic action.

Since the organizations of civil society originate primarily from communicative action, they are accurate reflections of needs experienced in the everyday lifeworld. These institutions are seen to work through the following process: Social and environmental issues that are created by social structures are experienced as burdens on everyday life (Habermas 1996: 365). These problems thus first become visible when they arise in the context of everyday life. These experiences can then lead to common concerns and support the formation of a narrative or discourse about their cause, nature, and resolution. These alternative discourses "find their concise expression in the languages of religion, art, and literature, the 'literary' public sphere in the broader sense, which is specialized for the articulation of values and world disclosure" (ibid.).

From their initial expression, these issues are discussed in the various media, in academic and professional organizations, and in other public forums. As the discourse becomes widely circulated, it "catalyzes the growth of social movements and new subcultures" (Habermas 1996: 381). Social-movement organizations form to represent groups of individuals dealing with unique situations in their lifeworld. They develop alternative discourses and practices, and present these social problems and their solutions in the public arena (ibid.). Thus, social movements can be conceived as "learning processes through which latently available structures of rationality are transposed into social practice—so that in the end they find an institutional embodiment" (Habermas 1979: 125). Thus, the institutions of civil society are seen as a key link in translating the impulses from the lifeworld into political demands for change (Habermas 1987a: 371–396; 1989a: 142). This institutional situation creates the possibility of civil society as the site for a restoration of democratic control over the political and economic systems (Habermas 1998b: 252).

That is not to say that civil society is the *only* site where social change can be developed. As Beck (1986: 23) has pointed out, the mass media, science, and law are also important institutions where democratic change would foster social learning. In addition, the administrative state is also a location in which attempts can be made to expand the range of alternatives considered in decision making. Actions by whistleblowers, committed activists, and contentious public administrators can foster the development of dialogue and democracy within this arena (Dryzek and Torgerson 1993: 136). However, without a robust social movement to lend political and legislative support to these efforts, little long-term change is likely to occur. Hence, one indispensable means of rapidly increasing our society's capacity for social learning is the development of strong, democratic social movements.

The Public Sphere

The effective translation of lifeworld concerns into political demands by the institutions of civil society requires that social institutions be responsive to democratically formed public opinion. The public sphere is the arena in which citizens have historically exerted their influence over collective decisions (Habermas 1962). Its role is to serve as an arena which can "subject persons or affairs to public reason, and to make political decisions subject to appeal before the court of public opinion" (Habermas 1989a: 141).

The public sphere constitutes a communicative structure formed by civil society and functions as an arena in which the institutions of civil society can identify problems, develop possible solutions, and create sufficient political pressure to have them addressed by constitutional governments (Habermas 1996: 359–360; 1998b: 248–249). The communicative link of the public sphere to the lifeworld via civil society "ensures that newly arising situations are connected up with existing conditions" (Habermas 1987a: 140). If proved successful through pragmatic action, new forms of behavior and concomitant worldviews can then become institutionalized. These new worldviews and social institutions can then successfully adapt our existing social institutions to changed conditions. This enables the realization of increased levels of rationality and social learning based on communicative action (Calhoun 1993: 392–393).

Since they are based in communicative action, the lifeworld and its representative institutions in civil society constitute a more rational, sensitive, and legitimate means to identify and resolve social and environmental problems than either market or state mechanisms (Habermas 1996: 381). These actions can then be translated through the public sphere to create effective political demand to structures that would be able to translate the impulses of the lifeworld into effective political demands to gain democratic control over the institutions of the market and the state (Habermas 1998b: 250).

The capability of a society to learn and respond to changed conditions is thus dependent on the generation of alternative world views, the open communication of these realities into the general stock of cultural knowledge, and the use of this knowledge in development of social institutions. To address ecological problems in our social order will require the development of open communication from a robust civil society in which environmental concerns are first perceived and identified by citizens, then accurately reflected and effectively presented by representative institutions of civil society, and then satisfactorily addressed by the political system. Thus, a key component of the development of an ecologically sustainable society is "radical and broadly effective democratization" (Habermas 1991).

Representative Organizations
To ensure a process of open communication from the lifeworld into the public sphere, the internal structure of organizations of civil society must be democratic and open. Hence, Habermas has concluded that the public sphere could be realized through the development of "organizations committed to the public sphere in their internal structure as well as in their

relations with the state and each other" (Habermas 1989a: 142). This leads Habermas to specify the requirements that these organizations must meet:

> To be able to satisfy these functions in the sense of democratic opinion and consensus formation their inner structure must first be organized in accord with the principle of publicity and must institutionally permit an intra-party or intra-association democracy—to allow for unhampered communication and public rational-critical debate. In addition, by making the internal affairs of the parties and special-interest associations public, the linkage between such an intra-organizational public sphere and the public sphere of the entire public has to be assured. Finally, the activities of the organizations themselves—their pressure on the state apparatus and their use of power against one another, as well as the manifold relations of dependency and of economic intertwining—need a far-reaching publicity. This would include, for instance, requiring that the organizations provide the public with information concerning the source and deployment of their financial means. (Habermas 1962: 209)

For Habermas, the structure of these institutions is not problematic. He concludes that they have "an egalitarian, open form of organization that mirrors essential features of the kind of communication around which they crystallize and to which they lend continuity and permanence" (1996: 367). As the following analysis will show, this situation is not pre-given in these associations.

While Habermas's analysis provides a comprehensive theoretical definition and justification for open and democratic social-movement organizations, its definition of the specific criteria of what constitutes a representative organization are vague. An examination of constitutional law can provide additional specifications of the criteria of an open and democratic social-movement organization.

In a series of three U.S. court cases, criteria for an authentic and representative organization has been developed and given binding legal status. The first case was *Sierra Club v. Morgan*, heard by the U.S. Supreme Court in the October 1971 term. In this case, the Sierra Club attempted to gain legal standing to prevent the development of Mineral King Valley in the Sierra Nevada. The Sierra Club, citing its "special interest in the conservation and the sound maintenance of the national parks," sought to bring suit over the proposed development. While recognizing the special interest and expertise of the Sierra Club in conservation, the Supreme Court held that the Sierra Club had not shown that the members that it represents were real parties in interest: "a mere 'interest in a problem,' no matter how long-standing the interest and no matter how qualified the organization is in evaluating the problem, is not sufficient by itself to render the organization 'adversely affected' or 'aggrieved' within the meaning of the Administrative Procedures Act" (405 US 727 1971: 645). Thus, the Sierra Club had no direct interest in the outcome, and its special interest and expertise did not entitle it to any special status in the development of Mineral King Valley.

The second case was *Hunt v. Washington Apple Advertising Commission*. The U.S. Supreme Court heard this case in the October 1976 term. The material question of this case was whether or not the Washington Apple Advertising Commission met the criteria to represent the Washington apple growers in a lawsuit. The Supreme Court examined the structure of the organization and held that the commission did have standing as an authentic representative of the apple growers. In their findings, the U.S. Supreme Court noted that the commission possessed "all of the indicia of membership in an organization. They (the apple growers) alone elect the members of the Commission; they alone may serve on the Commission; they alone finance its activities. In a very real sense, therefore, the Commission represents the State's growers and dealers and provides the means by which they express their collective views and protect their collective interests." (432 US 333 1976: 344–345)

These two cases were combined in *Health Research Group and Public Citizen v. Kennedy*. This case was heard in the U.S. District Court in Washington on March 13, 1979. At issue was the standing of Public Citizen, an organization founded by Ralph Nader, to sue the U.S. Food and Drug Administration over regulations governing some over-the-counter drugs. In the hearings, it was determined that a self-electing board of directors runs Public Citizen, that individual contributors fund it, and that its positions are "determined primarily by its employees." In spite of these characteristics, Public Citizen attempted to claim that it was a representative organization. Summarizing Public Citizen's argument in his findings, Judge John Sirica wrote:

Plaintiffs (Public Citizen) allege that their supporters and contributors influence Public Citizen's activities through their financial support and their letter writing. They also argue that the members of Public Citizen have a functional vote and have considerable influence on Public Citizen's policies and projects. Two factors—the individual contributions and the "continual communication" between Public Citizen and its contributors and supporters—are said to provide this influence. (82 FDR 21 1979: 27)

The Court rejected the arguments presented by Public Citizen. First, citing *Sierra Club v. Morton*, Judge Sirica maintained that standing cannot be gained merely by showing that an organization has an interest and specific expertise in an area. Control of the organization by its members is crucial, to ensure that the positions advocated by the organization are based on the interests of the affected party. Without such control, there is "simply no assurance that the party seeking judicial review represents the injured party, and not merely a well-informed point of view" (ibid.). Second, using the precedent set in *Hunt v. Washington Apple Advertising Commission*, Judge Sirica found that Public Citizen had no substantial nexus of control of the organization by its members, and hence it was not a representative organization. In coming to this finding, Judge Sirica defined the qualities of a representative organization. He maintained that to qualify as an authentic representative of its members an organization must have a "very substantial nexus between the organization and the parties it purports to represent." This means that the "members . . . normally exercise a substantial measure of power or control over an organization" (ibid.: 26). He also made a major distinction between a representative and a nonrepresentative organization: "In the Court's view, there is a material difference of both degree and substance between the control exercised by masses of contributors tending to give more or less money to an organization depending on its responsiveness to their interests, or though the expression of opinion in the letters of supports, on the one hand, and the control exercised by members of an organization as they regularly elect their governing body, on the other." (ibid.: 27)

Taken together, these three court hearings define the requirements of a representative organization (i.e., one that expresses the members' collective views and protects their collective interests). First, there must be a substantial nexus of control of the organization by the members. Second, the organization must be linked to people who have a direct stake in the outcome (i.e., who represent interested or affected parties). Conversely, this also defines what does not qualify as a "representative" organization.

Particular expertise or special interest in an area does not qualify an organization with standing. Thus, self-appointed "representatives," or experts, do not qualify as authentic representatives of collective interests. In addition, organizations without formal elections and mechanisms to enable membership control do not qualify as representative organizations. So-called informal mechanisms, such as informal communications, letter writing, or financial contributions, do not make an organization representative.

Conclusion

The development of our current social order can be seen as the expansion of the market and the state into more and more aspects of life. The progressive deterioration of the ability of the everyday lifeworld to affect the development of action has created a social order that systematically destroys the natural environment on which it based, as well as its social capacity for self-renewal.

Our current social formation not only expands into the lifeworld and disrupts the development and renewal of social order; it also creates our ecological crisis. Just as the invasion of the lifeworld by the market and the state progressively destroys the ability of our society to reproduce itself, the expropriation of the natural environment threatens to destroy the society's physical and biological basis. This restriction of communicative action restricts the learning capacity of the current social order. Without this self-corrective capacity, the destruction of the ecological basis of our social order cannot be addressed.

The associations of civil society can act to enhance the self-correction capability of our social institutions through creation of a robust civil society and restoration of the public sphere. Creating a robust civil society requires the development of democratic environmental movement organizations. By effectively communicating the imperatives of the lifeworld to the public sphere, the environmental movement's organizations could foster the development of a democratic and ecologically sustainable society. This would allow for the possibility of increasing the capacity for social learning, and it would foster the initiation of a restructuring of the current social order toward ecological sustainability.

One of the major ways to develop a strong environmental movement is a regeneration of environmental organizations along democratic lines (Goldblatt 1996: 202). By no means is this an easy project to accomplish. These associations are "vulnerable to the repression and exclusionary effects of unequally distributed social power, structural violence and systematically distorted communication" (Habermas 1996: 307–308). However, these associations also provide a valuable means of developing and advocating new social arrangements.

Habermas's theory of modernity provides criteria for evaluating the level of democracy in these organizations. The strength of this approach is that it specifies the requirements for a rational and representative political practice, and thus "opens up the possibility of linking normative considerations to empirical sociological ones" (Habermas 1992a: 448). Hence, one key to developing a solution to environmental problems lies in ecology-movement organizations that strive to recreate the public sphere in their internal structure. This theoretical analysis thus focuses on the "social level of institutionalized processes of deliberation and decision making" (ibid.: 461). We need to look at how the social movement's actual organizations are structured, and to evaluate how well they foster social learning.

4

Social Learning for Ecological Sustainability

If the lad or lass is among us who knows where the secret heart of this Growth-Monster is hidden, let them please tell us where to shoot the arrow that will slow it down.
—Gary Snyder (1990: 5)

Since environmental problems originate in the social order, their resolution requires social change. Though there have been many calls for society to change so that it does not continue to destroy the natural environment, how this process could be advanced politically remains vague. As a result, "the issue of social change is undertheorized in green politics" (Dobson 1993: 192). In essence, there is no plausible theory of transition to an ecologically sustainable society.

I have argued so far that communicative action in the lifeworld is crucial to renewing our social institutions so they will be able to meet new circumstances. The systematic blocking of communicative action has created a social order that is unable to institute sufficient social change to address continued ecological degradation. Restoration of control over social institutions by the lifeworld is a prerequisite for initiation of the actions through which environmental problems could be resolved. This links the resolution of ecological problems to the development of a democratic social order.

Organizations in the environmental movement are key actors in institutionalizing this process of social change. Their role is premised on their basis in the lifeworld, which enables them to debate and propose new social arrangements independent of the logics of capital accumulation and political power. By mobilizing citizens and by providing a competent, legitimate, authentic representation of their needs, environmental-movement organizations can help to bring about a democratic and ecologically sustainable society. As our current situation shows, these organizations have not been able to perform this task in the past. And it is not at all certain that they can accomplish this task now, or that they will be able to do so in the future. How can their ability to foster the creation of an ecologically rational society be improved?

The literature on the environmental movement does not provide an adequate foundation for an understanding of the overall movement, much less for an examination of its organizations' ability to foster social learning. Thus, there is a need to identify the conditions that either foster or inhibit an environmental organization's ability to communicate lifeworld concerns into the public sphere. To understand these conditions, and to develop a means to increase our society's social learning capacity, a close examination of the existing environmental organizations is needed (Habermas 1992a: 448; Offe 1992: 90; Stern, Young, and Druckman 1992: 145–146).

In this chapter I develop a framework for examining environmental organizations in the United States. The aim of this analysis is to develop an understanding of the factors that influence an environmental organization's capacity to foster social learning. This analysis could then assist in identifying the capabilities of these organizations, and perhaps it could inform action to remedy their limitations.

Social Movements, Social Change, and Distorted Communication

To conduct an examination of the existing environmental organizations as an aid to our understanding how social-movement organizations can succeed or fail in bringing about social change, we need to look at these organizations through three distinct theoretical frameworks: *frame analysis*, *resource mobilization*, and *historical analysis*.

Frame Analysis
Many of the attributes, capabilities, and limitations of a social-movement organization can be traced to the worldview or discourse on which the organization is based. The study of the worldviews of social movements has been formally developed in sociology by means of *frame analysis*. This approach focuses on the creation and maintenance of the common beliefs that define the reality in which a social movement exists.[1] At the center of

this type of analysis is the development of a movement's *discursive frame*—
the taken-for-granted reality in which the movement exists. The discursive
frame provides an interpretation of history that specifies the movement's
origins, its heroes, its development, and its agenda. This narrative gives an
organization its identity and guides its collective actions. It is a living and
contingent framework, created and changed by the participants in the
movement (Laclau and Mouffe 1985: 142–143).

The discursive frame of a social movement takes the form of a moral
drama in a quest for salvation in a new social order.[2] The world is seen as
a theater in which the drama of human life is played out. This drama
unfolds in a sequence in which the old social order falls into corruption.
Identification and elimination of the cause of evil follow. A new and
redeemed social order emerges, based on a new definition of reality that
then enters into competition with the dominant discourse (Stewart, Smith,
and Denton 1989: 152–155).

The first step in this process is the delegitimization of the dominant
worldview. This worldview is rejected when individuals feel that the dominant discourse is inadequate to the situation or when it becomes incoherent in the face of experience (Kuhn 1962: 93). This results in an expression
that takes the form of a rhetoric of discontinuity, which justifies a dramatic
change in society because of the problem situation and the need for action
(Jablonski 1980: 289; Griffith 1966: 460).

For a social movement to gain an identity, a new narrative of society must
then be created (Stewart 1980: 298–305). This voicing of a new representation of the social world is the act of identity formation of a social movement. This rhetorical process leads to the creation of a definable identity
for a social movement (Kurtz 1983). This frame contains four elements:

(1) diagnoses that identify problematic dimensions of power relations that are in
need of amelioration, (2) prognoses that articulate an alternative vision of power
arrangements, (3) compelling rationales for changing power relations and participating in movement drama, and (4) strategic and tactical direction delineating the
most effective means to obtain power. (Benford and Hunt 1992: 39)

This alternative discursive frame provides a basis for common interpretations of meanings and for the formation of collective actions. The discursive frame also links collective action to the individual through the provision
of justifications for joining the group. These justifications take the form of
a vocabulary of motives, which "supply adherents with compelling reasons

or rationales for taking action and provide participants with justifications for actions undertaken on behalf of the movement's goals" (Benford and Hunt 1992: 41).

A movement's discursive frame serves as a means of resocializing individuals into new identities. The discursive frame is linked to an individual's identity through reconstruction of the self based on the adoption of a new narrative. This new narrative redefines the past, the present, and the future, and reconstitutes the individual in a new symbolic reality. By providing this sequence, it serves as a means to restructure experience and constitute a new identity in the alternative worldview.[3] The acceptance of this new narrative of the self requires the establishment of resonance in the lifeworld of the individual.[4] The acceptance of this alternative social reality creates new social obligations. The level of commitment of the individual to support the social movement can vary from mild preference to total dedication. The level of acceptance is dependent on the level of personal exposure. As personal involvement with the specific situation grows in the lifeworld, the saliency of the discursive frame increases, and acceptance is more likely to occur (Gamson 1991, 1992; Melucci 1988; Eden 1993: 1752–1754). Thus, the discursive frame of a social movement defines the situation, places the interaction within a specific historical and social context, and provides a means of tying the individual's personal identity to collective action (Benford and Hunt 1992: 39).

The discursive frame also enables the construction and maintenance of movement organizations by constituting the social movement's shared version of reality. A discursive frame defines the nature of the problem and how the solution is to be achieved. This definition provides a basis for common interpretations of meanings, and the ability to coordinate action. The discursive frame also plays a role in the development of the internal characteristics of social-movement organizations. Since a particular discourse legitimizes certain actions and delegitimizes others, the strategy, tactics, and forms of resource mobilization of social movements are all related to the worldview on which a social movement bases its identity.[5] Out of this agreement, a network of interaction is formed which constitutes a social-movement organization (Brulle 1994; Emirbayer and Goodwin 1994: 1438–1449).

Once a movement has developed an identity, it moves into competition with other groups in society. This shifts the focus from the development of a movement's identity to the maintenance of this new definition of reality in relation to the dominant cultural beliefs. The emergence of a movement's identity occurs when the dominant and alternative worldviews clash. This creates the perception of a shared symbolic reality outside of the taken-for-granted social reality. This rhetorical competition creates a moral area of confrontation in which the existing situation is defined as morally intolerable by the alternative reality (Cathcart 1972, 1978, 1980). The challenging group redefines the situation as unjust and delegitimizes the existing authority. This new definition of the situation leads to collective action.

From the perspective of frame analysis, social-movement organizations are seen as the creation and institutionalization of a historically developed discursive frame which provides an identity to specific networks of collective action (Benford and Hunt 1992; Brulle 1996; Touraine 1977). In conducting an analysis of a complex social movement, frame analysis makes no assumption that there is any fundamental structure or framework to the different discourses. Instead, there are multiple particular discourses that are the historical constructions of social actors. The objective of the analysis is to describe the multiple discourses that have been defined by the actors themselves. The goal is to embrace the particular and historical diversity of beliefs. Accordingly, "plurality is not the phenomena to be explained, but the starting point of the analysis" (Laclau and Mouffe 1985: 140).

Discursive Frame and Distorted Communications

The discursive frame has a major influence on the capabilities and limitations of an environmental organization. First, it can limit the range of alternatives considered, and the level of political appeal. A movement's discursive frame creates a binding definition of the situation. Accordingly, this frame enables certain aspects of the world to be seen, and excludes others. Thus, the effect of developing a discursive frame is the exclusion of alternative realities. This discourse can illuminate or mask consideration of the causes of ecological degradation. When it obscures or limits consideration of other alternatives, it can limit the range of options considered. It can also fail to connect with larger audiences as a result of its limited range of concern. This can limit the identification of sources of solidarity, the range of alternatives considered, and the possible means of resolving environmental problems. The failure of a discourse to successfully link its specific appeals to the definition of the larger public-good results can be seen as a failure of the discourse. As results of such a failure, "public divisions are petrified, conflicts are prolonged and solutions are deferred" (Killingsworth and Palmer 1992: 8).

In addition to limiting the range of alternatives debated, a discursive frame may foster the development of oligarchic or democratic social-movement organizations. In his study of the determinants of the governmental structure of social-movement organizations, Knoke (1990) examined the determinants of power distribution inside various social-movement organizations. His analysis shows that the cultural perspective of the institution outweighed any other factor as a determinate of organizational structure: ". . . governance may be less a matter of structural or strategic contingency and more a consequence of cultural institutionalization" (ibid.: 160). This position was amplified by Gamson (1991), who showed that one of the most important influences on the internal structure of an organization is the discursive frame in which the social movement forms an identity. Since the movement's frame defines the organization's identity, inclusion of democratic principles into the frame serves as a powerful barrier to development of an oligarchic structure. Schmitter (1983: 918) also noted the influence of the discursive frame on organizational structure: ". . . many of these groups define their very existence in ways that defy professionalized representation and bureaucratic encadrement. To be organized corporatistically would destroy the very basis of their collective identity." Thus, a social movement's discursive frame can be an important determinant of the organization's structure.

Resource Mobilization

In addition to its discursive frame, one of the major influences on the capacities of a movement organization is the role played by material resources in either facilitating or discouraging the development of group action. The sociological perspective of resource mobilization examines this process. At the core of resource mobilization are a number of shared assumptions (Jenkins 1983; Morris and Herring 1987):

• Social movements must be understood in terms of conflict between social movements and existing institutionalized polity.
• Conflicts of interests are preexisting and are based on the structural and power relations in a given society.

• The basic goals and identities of movements are determined by conflicts of interest.

• Emergence of social movements is brought about by changes in material resources, group organization, and opportunities for collective action.

• Social movements typically take the form of large, centralized, formal organizations.

• The success of a movement is evidenced by recognition of the group as a legitimate member of the institutionalized polity, and/or by increased material benefits.

• Determinants of success include degree of formal organization, level of repression by elites, availability of selective incentives to join the movement, the resource-mobilization capacity of the existing polity, and willingness to use "unruly" methods.

There are two key approaches that can provide an understanding of the characteristics of environmental-movement organizations. They are divided between models of the internal and the external dynamics of movement organizations.

Internal Dynamics

The perspective on the internal dynamics of movement organizations originated in the work of Robert Michels. For Michels, a political organization followed an inevitable pattern of development from democracy to oligarchy in a cyclic rotation of elites:

The democratic currents of history resemble successive waves. . . . When democracies have gained a certain stage of development, they undergo a gradual transformation, adopting the aristocratic spirit, and in many cases also the aristocratic norms, against which at the outset they struggled so fiercely. Now new accusers arise to denounce the traitors; after an era of glorious combats and of inglorious power, they end by fusing with the old dominant class; whereupon once more they are in their turn attacked by fresh opponents who appeal to the name of democracy. It is probable that this cruel game will continue without end. (1915/1962: 371)

This approach was summarized in Michels's Iron Law of Oligarchy: "Who says organization, says oligarchy." (ibid.: 365)

This perspective of the transformation of organizations into oligarchies was also reflected in the work of Max Weber. For Weber, the creation of organizations is based on the charisma of the leader. However, when the leadership is passed to the next generation, a process known as "the routinization of charisma" occurs. The formation of a stable organization requires the transformation of personal authority to an organizational authority (Weber 1918/1978: 246). This results in the formation of a bureaucracy. This change in the form of authority also results in a change in the interests of the administrative staff, away from social change, and toward ensuring their positions in the organization (ibid.: 247). This process is summarized in the Weber-Michels model[6]:

As a movement organization attains an economic and social base in the society, as the original charismatic leadership is replaced, a bureaucratic structure emerges and a general accommodation to the society occurs. The participants in this structure have a stake in preserving the organization, regardless of its ability to attain goals. (Zald and Garner 1987: 121)

This model defines three components of this shift: a transformation of goals from utopian to more pragmatic ones in line with social consensus; a shift to organizational maintenance, which involves ensuring the flow of funds and maintenance of membership; and a transition to an oligarchic structure as a formal bureaucracy develops (Zald and Garner 1987: 121–122; Offe 1990: 239–240).

Resource-mobilization theorists elaborated this model further. In their 1956 book *Union Democracy*, Lipset et al. expanded on the work of Weber. Their analysis showed that a number of specific criteria would determine the level of internal bureaucratization or democracy within an organization. Further research showed that the shift toward bureaucracy hypothesized by the Weber-Michels model was not inevitable. Instead, the internal characteristics of an organization were seen to be functions of a number of variables (Jenkins 1983: 541; Zald and Garner 1987: 121–142).

Among the key variables determining the internal capabilities of a movement organization are its means of recruiting members and its methods of raising funds. The sociological models of recruitment to a social-movement organization are typically based on face-to-face interaction in "grassroots" organizations whose members are recruited through personal contact and lifeworld communication. However, grassroots organizations are being supplanted by mass organizations that rely on mass mailing. Though this may be important for fund raising and thus for organizational maintenance, it can reduce an organization's capacity for political mobilization, isolate its members from its leaders, and consequently limit the capacity of the organization to initiate social change. Thus, it provides an example of the importance of institutional maintenance over goal achievement.

In a mass organization, there is little or no member involvement (Hayes 1986). Instead, the membership consists mostly of "checkbook" members, affiliated with the group only by virtue of monetary contributions solicited via mass mailing. There is virtually no face-to-face contact with other members in a common setting. Many of these groups are staff organizations, lacking any real membership base and having an oligarchic, nonparticipatory structure. To the extent that there is membership involvement, it is marginal, and it consists basically of making economic contributions to the group. This creates a group with only fleeting and shallow membership support, totally dependent on the remote and impersonal national leadership of the group and thus vulnerable to manipulation (ibid.: 135). The end result, according to Herndl and Brown (1996: 9), is that "the possibilities for participation and rational engagement are radically limited. . . . This limitation, perhaps one largely imposed by the medium, reduces the process of ethical decision making so severely that it essentially guts any viable sense of collective, community-based ethical action." Thus, a mass-mailing-based group limits the development of individual commitments to the alternative discourse through participation. It also tends to concentrate power in the central staff of the organization, and makes the so-called members into contributors without any knowledge of the actions of the organization other than what is available from the self-serving publications of the group itself.

External Dynamics

The second perspective on social-movement organizations focuses on the interaction between the challenging group and the established institutions in the social order (Tilly 1978, 1985; Gamson 1975; Oberschall 1978). The actions of external institutions have major effects on the creation and the maintenance of social-movement organizations (Neidhardt and Rucht 1991: 443–445).

The traditional concept of founding and maintaining a social-movement organization is based on idea of mobilizing individuals into a group capable of collective action. This group is nominally based on face-to-face interaction (Hayes 1986: 134–135). The structure of the group is seen to derive from the member's communicative action. However, the state and private institutions play an active role in the creation of social-movement organizations. First, a network of institutional requirements formed by government and private organizations limits the development of social-movement organizations in the United States. Government actions that restrict the form of social movements include organizational requirements to obtain and maintain nonprofit income-tax status and requirements to qualify to publicly solicit for charitable contributions. Private charitable rules also limit organizational activities. For example, combined charity appeals have a series of organizational requirements that must be met to be included in their list of approved organizations, and organizations that monitor charitable giving have developed a number of screening criteria for acceptable organizations. Second, social-movement organizations can be created and maintained through a top-down rather than a grassroots process (Walker 1991: 186). The key component in this elite development of social-movement organizations is financial patronage (Walker 1983, 1991). The primary government economic supports are outright grants, contracts for services, and postal-rate subsidies (McCarthy et al. 1991; Walker 1991: 32–33).

External Funding and Social-Movement Organizations
Financial support provided by private foundations and individual patrons exerts a powerful influence on the capabilities of social-movement organizations (Walker 1991). The key external institutions that provide funding to movement organizations are government agencies, corporations, and foundations. External funding creates a dynamic that can be seen as financial steering of the movement organization. This process has been extensively studied in the literature on the influence of foundations on movement organizations. However, this dynamic of financial steering is not limited to foundations. A similar dynamic operates with regard to the influence of corporations and government agencies on movement organizations.

Foundations are "part of the corporate community in terms of their interorganizational affiliations" (Salzman and Domhoff 1983: 208), and they constitute a system of power and influence (Roelofs 1986: 1; Colwell 1993: 3–4; Nielsen 1972: 406). Thus, private foundations are "institutionalized agencies of the capitalist class" (Jenkins and Eckert 1986: 819). As such, they are not guided by linguistic media; rather, they are located inside the logics of capital accumulation and political power. Private foundations constitute a "system of control largely unacceptable to citizens"

(Roelofs 1986: 2), and they serve as a systematic source of the distortion of communication within social-movement organizations.

Private foundations gain their influence over social-movement organizations through their financial power. Foundation funding plays a major role in the initiation and the maintenance of movement organizations (Jenkins 1987: 299). Jack Walker has documented the importance of outside funding for the startup of citizen groups. Based on a 1980 survey of 115 citizens' groups, Walker (1991: 78) found that 89 percent of the groups were founded with outside financial aid. In addition, foundations are not passive actors selecting among proposals. Foundations have increasingly taken a more activist role in the development of movement organizations, including formation of their own organizations (Ylvisaker 1987: 363). Outside funding increased steadily throughout the twentieth century. Walker compiled data on funding sources by time period in which the organization was founded. Reviewing this shift in funding, Walker (ibid.: 49) concludes that one of the major reasons for the rise of movement organizations in the 1960s and the 1970s was "the emergence of many new patrons of political action who were willing to support efforts at social reform."

In addition to initiation of new social-movement organizations, foundation funding plays an important role in maintaining existing organizations. Walker (ibid.: 82) reports that membership dues constitute only 32.7 percent of the income of citizens' groups. Outside sources, including government contracts and grants, corporate gifts, foundation grants, and large gifts from individuals and other associations, totaled 38 percent of the annual income of 206 citizens' groups. Based on this analysis, Walker concludes that "without the influence of patrons of political action, the flourishing system of interest groups in the United States would be much smaller and would include very few groups seeking to obtain broad collective or public goods" (ibid.: 101–102).

Foundation funding draws social-movement organizations into a network of power and control. This limits the range of organizational forms and goals (McCarthy et al. 1991: 69–70). One means of control is the establishment of criteria for combined charity appeals. Failure to conform to these criteria can be a significant limitation in obtaining charitable or foundation funding for a movement organization. In addition, association with a foundation enhances the social status and legitimacy of an environmental organization. In discussing the advantages of association with a large foundation for an environmental organization, Donald Snow (1992: 65) maintains that this relationship not only increases the organization's overall prestige but also enhances its power:

> Leaders depend on foundations for programmatic ideas, information on productive networking with other groups, occasional technical support, and the sense of legitimacy and prestige that comes with foundation grants. Well-funded organizations gain the attention of policymakers simply by virtue of the recognition they receive from national grant makers.

Most important, however, foundation funding allows a foundation to gain direct influence over a movement organization through participation in its board of directors (Colwell 1993: 105). The funding of movement organizations by foundations shifts the locus of control of the organization from the membership to the financial patrons (Hayes 1986: 143; Schmitter 1983; Wilson 1990: 70; Jenkins 1989: 311). This reduces the independence of the organization and changes its organizational form, its practices, its efforts at grassroots mobilization, and its legitimacy.

Shift in Organizational Form
Foundation funding generally brings about a shift in the management practices and form of a movement organization. Two general requirements for obtaining foundation funding are a centralized accounting system and executive management of program activities. A movement organization may adopt this form of management in order to gain access to funds (Jenkins 1987: 307). This centralizes the organization, and it "can disenfranchise broad-based participants and lead to elite dominance" (Powell and Friedkin 1987: 191). In addition, the development of technical capabilities expands the role of an organization's staff. As a result, participation in the organization declines, and the control of staff over the organization increases (ibid.: 191).

Change of Organization's Practices
Foundation funding has the capability to change a social-movement organization's strategy toward advocating noncontroversial positions and engaging in traditional and nonconfrontational practices for social change; it also fosters corporatistic decision making. To obtain funding, movement

organizations must undertake projects for which grant support can be obtained. Research shows that private foundations are "generally . . . politically cautious in their support for social reform" (Jenkins and Eckert 1986: 819). They generally fund social movements that undertake noncontroversial projects. As Colwell (1993: 4) notes, "activist groups or researchers who challenge current political and economic arrangements are generally not funded." Thus, in order to obtain funding, movement organizations may adopt more conservative and politically acceptable positions.

In addition, foundation funding exerts an influence to shift the practices of a movement organization from "outsider" strategies of disruption to more conventional political action (Jenkins and Eckert 1986: 828; Walker 1991: 120; Powell and Friedkin 1987: 191). Organizations that continue to engage in confrontational tactics are likely to have their funding removed (Jenkins 1987: 301). The net result of this process is "a narrowing and taming of the potential for broad dissent" (McCarthy et al. 1991: 69–70).

Finally, foundation funding encourages the development of a neocorporatist political practice that restricts the power of a social-movement organization to act independently. This ties a movement organization into a network of financial patrons, and it shifts control from the members to the patrons. The organization becomes dependent on a network of actors dominated by large foundations and major contributors (Schmitter 1983; Hayes 1986; Wilson 1990, 1991). The result of this is the development of neocorporatist politics. In summarizing his analysis of foundation influence on movement organizations, Jenkins (1989: 311) maintains that "foundation patronage . . . could be interpreted as creating a neocorporatist system of political representation in which elites exert increasing control over the representation of social interests."

Reduction of Grassroots Mobilization Efforts

The maintenance of a social-movement organization is dependent on a steady flow of economic resources. These resources can be supplied either by an active membership or through external support. Foundation funding can relieve a movement organization of the continued effort to mobilize members and to obtain their financial support. Thus, it can lead to the decline of efforts to mobilize grassroots support, indirectly reducing the organization's political strength (Jenkins and Halcli 1996: 26).

Loss of Legitimacy

Foundation funding can create and maintain social-movement organizations with very limited or no ties to their purported membership. This can serve to undermine the legitimacy of these organizations in the public sphere (Habermas 1996: 352). These movement organizations may have excellent technical and legal arguments. However, lacking membership participation, and controlled by patrons and staff, they are vulnerable to the charge of elitism. This reduces their legitimacy in public forums (Jenkins 1987: 308).

Sponsorship by external institutions creates a social-movement sector with two distinct spheres. The first sphere comprises indigenous movement groups, based in face-to-face mobilization and deriving most of their resources from their members. The second sphere comprises professional organizations, controlled by staffs and boards and based in external funding and mass mailing, that purport to speak for their supposed beneficiaries (Colwell 1993: 199).

The impact of external organizations on movement organizations can range from channeling to co-optation. Channeling refers to the indirect and unintended consequences of foundation sponsorship, including centralization, professionalization, and the tendency of movement organizers to undertake activities that will please their sponsors (Roelofs 1986: 9). Co-optation refers to a process of social control of a movement organization by a foundation through representation on the board of directors, directed grants, or creation of foundation-based organizations. The extent of these two processes cannot be determined without detailed analysis of a specific relationship.

Whether or not a particular foundation channels or controls a particular movement organization, external funding distorts the public space by placing movement organizations under the influence of external organizations. It also gives certain movement organizations a larger voice in the public arena than other movements. In essence, some discursive frames become more highly represented in the public space than their citizen support would allow. "By funding professional organizations," Jenkins (1989: 311) notes, "foundations (and government elites) strengthen the political advocacy of a limited number of centrally controlled organizations." External funding thus can serve as a systematic means of distorting communication in the public sphere.

Historical Analysis

The third form of analysis of social movements is historical and sociological analysis of patterns in collective behavior. Social-movement organizations are formed in response to general social patterns. However, they are also formed in particular historical circumstances. Thus, to understand their capacities and limitations, it is not enough to examine them through sociological theories. A sociologically informed historical analysis of the development of social change is a prerequisite to understanding how current environmental organizations developed and how much potential they have for future action. The full force of contingent historical events must be accounted for. In short, history counts.

Historical analysis plays an important role in understanding the goals and practices of social movements. Contingent historical events have a major impact on what is now occurring and on the possibilities of future action. According to Worster (1994: 8), "all ideas, past and present, are grounded in particular historical contexts." Thus, to understand the goals, capabilities, and limitations of existing environmental organizations we need to understand the situation in which the ideas that define these organizations originated and developed.

Furthermore, the historical process serves to inform what possible future actions can be taken. Tilly (1981: 215) explains: "The accounts require historical grounding, most obviously, because the known means of action that are available to people (a) vary significantly as a cumulative product of historical experience; and (b) strongly constrain the likelihood and character of collective action." By examining the historical development of the environmental movement, an understanding of what practices have led to success and what practices have failed can be gained. This can expand the collective action repertoire of cultural knowledge of the environmental movement, and perhaps lead to more effective actions.

Historical analysis of the environmental movement's organizations requires empirical investigation of the discursive frames, the particular practices, and the events that created and shaped this movement. This analysis is informed by sociological theory by specifying what events we should examine to understand the process of social change. It also adds the dimension of time to the analysis of social phenomena (Tilly 1981: 44). Through this process, the static, ahistorical abstractions of sociological theory can be transformed into accounts of the origins and the development of the current environmental organizations.

Two excellent examples of this form of analysis are *The Social Origins of Dictatorship and Democracy* (Moore 1966) and *The Making of the English Working Class* (Thompson 1963). Moore separates the development of democracy from modernization and offers historically contingent explanations of the presence or absence of democratic institutions based on a comparative study of the actions of groups in several different societies. Thompson shows how the English working class acted together to make history. Rather than the inevitable working out of a Marxist dynamic, Thompson shows how the working class acted collectively to forge itself into a historically consequential, class-conscious proletariat.

The three approaches reviewed in this section can inform an analysis of the ability of social-movement organizations to foster social learning and of their effectiveness in doing so. Frame analysis suggests that certain discourses can mask consideration of the causes of ecological degradation and thus limit the range of options considered. In addition, the discursive frame influences the structure of an organization. Thus, some discourses may foster the development of oligarchic institutions. Furthermore, resource-mobilization theory suggests that the means and sources of funding can have an important impact on the independence and management of a social-movement organization. Finally, historical analysis is needed to identify the particular circumstances in which the organization exists.

Grassroots or Astroturf?

One way to visualize the extent of distorted communications within environmental organizations is to construct a dichotomy between a "grassroots" and an "Astroturf" organizational structure. The term "grassroots" can, at minimum, be taken to define "an essentially democratic, locally based" group (Gould et al. 1996: 3). However, a more robust definition can be developed to understand what would truly constitute a democratic grassroots organization.

The theoretical criteria developed from the theory of communicative action and the legal definitions provided by U.S. constitutional law converge and compliment one another. Together, they define a representative or grassroots organization. A grassroots organization is an authentic and legitimate representative of community that is based on open and undistorted communication in the lifeworld, and a substantial nexus of control of the organization by the members exists through a formally specified democratic organizational structure. The policies adopted and advocated by a grassroots group are developed from communicative action, and their legitimacy is based on the reasoned acceptance of the membership of the group. The term "grassroots" thus conveys the idea of an organic and living organization with its roots in the lifeworld of the community.

An "Astroturf" organization, though it may look like a grassroots group, is a synthetic creation made of artificial material and having no organic link to the community. It is a nonrepresentative organization that could not pass the legal criteria of a representative one. It is not an authentic and legitimate representative of a community. It is based in economic or political sponsorship and/or mass mailings. It has a closed and nonparticipatory structure, and no substantial nexus of control of the organization by the members exists. The policies adopted and advocated by an Astroturf group are developed by the staff and the board of directors, and their acceptance by the members is based on mass communication via advertising or direct mail. Thus, an Astroturf organization is not an authentic representative of a community. Such organizations compromise the independence of a movement, and they are susceptible to subordination and control by the market and the bureaucratic state. Instead of standing outside these two systems of instrumental reason (the market and the state), "Astroturf" organizations become co-opted and controlled by the logic of money or that of political power. This limits both the development of democratic control over the political and economic systems and the potential for social learning. Thus, it reduces our society's ability to resolve the problem of environmental degradation.

From the perspective of these two ideal types of organizational structures, we can begin to examine the organizational structure of the environmental movement. The reality is much more complex than this simplified dichotomy. It is seldom the case that a group is purely a grassroots or an Astroturf organization. There are degrees of organizational democracy and representation. It is more useful to see Astroturf and grassroots organizations as defining two ends of a spectrum of forms of organizational structures. Thus, a more nuanced definition of organizational structure is needed.

There are four types of organizational structures that can be used to identify the extent of organizational democracy:

Oligarchy   The organization is governed by a board of directors, which elects the officers and the new board members. No provisions for input by individual members exist in the by-laws.

Representative   Members of the organization can elect representatives of their local chapters. These representatives then participate in the selection of the organization's board of directors, officers, and policies.

Limited Democracy   The organization is governed in part by a board of directors and in part by the members. Individual members can nominate and/or elect some of the members of the board or some of the officers. However, certain aspects of organizational control are specifically delegated to the board of directors.

Democracy   The organization is governed by its members. The board of directors and the officers are nominated and elected by the membership. Policies can be debated and voted upon by individual members.

As I have argued in this section, mobilization of citizens to create political demand for change can give way to solicitation of funds via mass mailing. In addition, as external patronage and sponsorship increase, an organization is increasingly co-opted and controlled by means of money or political power. In addition, the environmental movement's discursive frame may limit the range of alternatives debated and may foster the development of oligarchic or democratic organizations. Thus, instead of representing the lifeworld, environmental organizations can become isolated, unrepresentative, and subordinated to control by the market and the bureaucratic state. This limits the potential development of a democratic public sphere and the potential for social learning.

A Framework for Examining Environmental Organizations

To identify the social factors that can assist in developing a democratic and sustainable society requires research based on the combined perspectives of historical development, frame analysis, and resource mobilization.

The initiation of this task needs to first take into account the existing studies of the U.S. environmental movement. These studies can be divided into three categories.

Existing Studies of Environmental Organizations

Those in the first category analyze the movement's cultural models in various ways.[7] Environmental belief models are often based on the problematic use of either essentialist or structural linguistic forms of analysis.[8] In these forms of analysis, the social scientist, instead of seeing the discursive frames as contingent structures, imputes some core structure that underlies the surface beliefs.[9]

Linguistic structuralism has been robustly criticized as being arbitrary, and as being reductive because it replaces the complex of meaning that is constantly shaped and changed in communicative action with an abstract construction of the social scientist (Bleicher 1982: 88–104; Bottomore 1993: 649; Norris 1982; Simons 1990; Norris 1982: 52). Examining the work of Roderick Nash and Samuel Hays, Donald Worster (1985a) critiques the use of structuralist models in constructing a history of environmental beliefs. Worster maintains that the classification schemes used by Nash and Hays are based on an uncritical appropriation of discredited modernist and essentialist models. Accordingly, he concludes that their use of structuralist and teleological models is "a suspect strategy" (ibid.: 258). Worster has argued (1977: 346–347) for rejecting such simplified explanatory schemes and embracing the complexity and diversity of the environmental movement.

The second form of analysis is qualitative or quantitative examination of organizations in specific sectors of the environmental movement. Analyses of this type provide in-depth examination of the discourses and practices of a subset of the movement, or aggregate patterns of several major environmental organizations.[10] These types of studies provide several key insights into the movement's organizations. However, there are significant limitations. For one thing, because of the self-limited nature of these studies, no collective image of the organizations that constitute the environmental movement emerges. The existing literature is limited to studies of specific (usually radical) groups and analyses of the largest and most visible organizations. Hence, there is no overall study of the U.S. environmental movement's range of beliefs or organizations. In addition, though there have been some attempts to place these studies of group dynamics into a historical perspective (e.g., Dunlap and Mertig 1992), none of the organizational studies are connected to previously described theories linking these organizations to social change (Catton 1980; Catton and Dunlap 1978, 1980; Schnaiberg 1980). Finally, none of these studies measure the amounts of resources mobilized by the various organizations (Brulle 1994).

The third way to study the environmental movement is through the use of New Social Movements (NSM) theory.[11] This approach traces the development of "new" social movements to changes in the structure of modern society, the development of alternative discourses, and formation of collective action based on these discourses.

The movements to which NSM theory have been applied are seen as "new" in respect to the labor movement, and as having specific characteristics that distinguish them from that movement. They are generally not concerned about the distribution of the material production of the society; rather, they are concerned with what constitutes the "good life," or with identity issues. Their economic support is not based in class conflict; rather, they are supported primarily by highly educated workers in the service sector or by the marginally employed and the unemployed. Participation in NSMs is motivated by altruistic values rather than by economic self-interest. NSMs are postulated to have democratic organizational structures and pluralistic politics. NSMs are supposedly characterized by political activity outside the traditional left-right political-party structure of Western democracies. And NSMs use nontraditional and unconventional means to forward their goals (Dalton, Kuechler, and Bürklin 1990: 14–16).

This perspective has been subjected to some telling criticisms.

First, the distinction between "old" and "new" social movements is overdrawn (Ray 1993: 61; Calhoun 1994: 180–183). The alleged newness of certain social movements is based on a distorted notion of the labor movement that ignores the richness and the varied nature of workers' movements. In addition, other historical social movements of the nineteenth century and the early twentieth century, such as the women's movement and a variety of religious movements, are ignored. In his examination of NSM theory, Larry Ray (1993: 61) argues that "claims to novelty are exaggerated and ahistorical since most contemporary movements have long histories, and movements like environmentalism, pacifism, and feminism, were significant around 1890–1900 or before."

Second, NSM theory lacks a connection to the existing research on social movements. The NSM theories lack adequate grounding in the historical, empirical, and discursive research on particular movements, such as the environmental and women's movements (Tucker 1991; Kaase 1990: 98). Much of the NSM literature reads as like projection of desired character-istics onto the subject matter, not like actual analyses of existing social movements. The lack of evidence has been noted by Rucht (1989: 63): "Though the decentralized network structures of new social movements have been often emphasized, there is no clear evidence that such structures are always present. . . . Furthermore, there is a risk that the self-image of new social movements as grass-roots organizations can be too easily adopted by external observers." Examining the work of one of the lead-ing NSM theorists, Kenneth Tucker (1991: 84) argues that Jean Cohen "neglects any true historical grounding for her theories" and "tends to make broad, sweeping statements, conflating social thought, social move-ments, and historical context."

Finally, the generalizations of the NSM perspective miss several key fea-tures of the development of the environmental movement. The clear his-torical periodization in the development of different environmental discourses is neglected. The role of the expansion of environmental degra-dation in the rise of the movement is not considered. Goldblatt (1996: 133) concludes that "this focus generates a model of cultural change that is unsuited to historical explanation or to assessing the role of cultural ideas and values in the mobilization of environmental politics."

The analyses of environmental-movement organizations in the NSM lit-erature are thin. Studies of discursive frames are limited, partial, and often based on flawed models. While there are several excellent studies of seg-ments of the environmental movement, no overall image of the movement becomes visible. Finally, the NSM literature offers a theoretical model with-out an empirical base.

Thus, the existing literature on the environmental movement needs to be expanded in three areas: analysis of the dimensions of the movement in the United States, identification of the discourses on which the components of the movement are based, and details of the organizational practices that each discourse enables.

Dimensions of the Movement

The first task of the analysis is to measure the environmental movement's resource-mobilization capacity. The movement's ability to re-create the public sphere is partially based on its overall influence in society. One aspect of this influence is the movement's ability to mobilize sufficient resources over a sustained period of time. The key issue is whether or not the movement has sufficient resources to be a substantial social actor. Analysis of the various components of the environmental movement has been a continuing topic of a vast number of articles and books over the past 20 years. In these studies, it is generally assumed that environmentalism is a significant social actor in the political arena. However, there has been no real analysis of the resources that are mobilized by this social movement. As a consequence, the theoretical perspectives developed by resource-mobilization theories regarding the importance and influence of the environmental movement remain untested.

The Historical Development of the Discursive Frames of the Environmental Movement

The second task of the analysis is to identify the discourses on which the components of the environmental movement are based. An analysis of the relationship between the development of discursive frames and formation of social-movement organizations can facilitate a critical understanding of the ability of the environmental movement to foster social learning. One way to understand the complexity of the U.S. environmental movement is to see it as a series of historically developed networks of communicative action based on multiple discursive frames (Brulle 1992, 1996; Eder 1996_; Neidhardt and Rucht 1991: 446).[12] These discursive frames form the basis for the many different forms of action, organization, and objectives within the current movement.

The U.S. environmental movement is composed of several thousand organizations, including the Sierra Club, Greenpeace, Earth First!, and the Citizens' Clearinghouse for Hazardous Waste. These groups have a wide range of objectives, strategies, and internal structures. The first of them were founded more than 150 years ago. As the environmental discourses developed into organizations, they created specific communities. Because they are concerned with different forms of ecological degradation, these communities' discursive frames focus on the many different components of the overall environmental problem (Schlosberg 1998: 21).

Killingsworth and Palmer (1992: 18) note that "as the environmental crisis has intensified, the identities of special interest groups and their corresponding rhetorics have tended to undergo strong dislocations, with new groups emerging from old ones to represent each new stage of a deepening political consciousness." "The older groups do not, however, just die off," Killingsworth and Palmer point out; "they remain in existence as points of tension for the new groups."

The fact that its various organizations originated in very different historical circumstances explains why the current U.S. environmental movement is made up of numerous, partially overlapping communities. As a result of this process, "there is no single agenda or integrated political philosophy binding the hundreds of environmental movement organizations found in the nation" (Bullard and Wright 1992: 41). Indeed, the development of these organizations is the outcome of contingent historical events, the development of specific discourses, and mobilization of material resources to create these organizations. The unifying thread that can connect historical events with the development of the environmental movement's organizations is the discourse that forms the identity of this social movement. The discourse of a movement translates the historical conditions and the potential for mobilization into a reality that frames an organization's identity. This identity then influences the organization's structure, tactics, and methods of resource mobilization.[13]

An analysis of U.S. environmental organizations from a discursive perspective focuses on the historical creation of the various discourses and their subsequent instantiation into organizations with definable identities. The discursive frames that define the components of the U.S. environmental movement are listed in table 4.1. It is important to note that this scheme is not an arbitrary or abstract construction. Instead, it is an initial attempt to define existing communities of action that are based on common discourses. These discourses are not static or reified structures, nor are they essentialist models that exist apart from the practices of interaction. Rather, this scheme seeks to adequately typify the orientations through which the actors define their communities.

Table 4.1
Discourses of the U.S. environmental movement.

Manifest Destiny
The natural environment is unproductive and valueless without development. Hence, the exploitation and development of abundant natural resources for economic development contributes directly to human welfare.

Wildlife Management
The scientific management of ecosystems can ensure stable populations of wildlife. This wildlife population can be seen as a crop from which excess populations can be sustainably harvested in accordance with the ecological limitations of a given area. This excess wildlife population can be used for human recreation in sport hunting.

Conservation
Natural resources should be technically managed from a utilitarian perspective to realize the greatest good for the greatest number of people over the longest period of time.

Preservation
Nature is an important component in supporting both the physical and spiritual life of humans. Hence, the continued existence of wilderness and wildlife, undisturbed by human action, is necessary.

Reform Environmentalism
Human health is linked to ecosystem conditions. To maintain a healthy human society, ecologically responsible actions are necessary. These actions can be developed and implemented through the use of the natural sciences.

Deep Ecology
The richness and diversity of all life on earth has intrinsic value, and so human life is privileged only to the extent of satisfying vital needs. Maintenance of the diversity of life on earth mandates a decrease in human impacts on the natural environment and substantial increases in the wilderness areas of the globe.

Environmental Justice
Ecological problems occur because of the structure of society and the imperatives this structure creates for the continued exploitation of nature. Hence, the resolution of environmental problems requires fundamental social change.

Ecofeminism
Ecosystem abuse is rooted in androcentric concepts and institutions. Relations of complementarily rather than superiority between culture and nature, between humans and nonhumans, and between males and females are needed to resolve the conflict between the human and natural worlds.

Ecotheology
Nature is God's creation, and humanity has a moral obligation to keep and tend it. Hence, natural and unpolluted ecosystems and biodiversity need to be preserved.

The selection of these nine discourses was based on study of environmental discourse in the field of environmental philosophy (where there is a building consensus on this topic) and on hermeneutic analysis of the different environmental discourses.[14] The hermeneutic analysis in chapters 6–9 below proceeds through examination of the specific historical events and texts involved in the development of each discursive frame and the subsequent creation of movement organizations. The plausibility of this scheme is based on a demonstration of its ability to integrate a wide variety of literature on the U.S. environmental movement from a variety of sources (including environmental sociology, philosophy, and history) with textual evidence from the founding documents of paradigmatic movement organizations.

Discursive Frame and Organizational Practices

The third task involves analysis of the organizational practices that each discourse enables. Discourse analysis can contribute to the understanding of the practices of movement organizations, including their internal structures, strategies for social change, and resource-mobilization practices. Combining the discursive frame with an analysis of the practices of movement organizations makes possible an examination of the factors that either foster or restrict communication within these organizations.

Analysis of discursive frames can show how it is possible for movement organizations to enter into discourse traps and ideologies that dichotomize the causes of ecological degradation, limit and channel debate into ideological contests, and restrict the nature of the options that can be considered in the search for solutions (Killingsworth and Palmer 1992: 9–10). A cultural critique of the limitations of the discourses can assist us in identifying these characteristics in the various ecological discourses and in developing attempts to avoid these traps.

This research can also shed light on the conditions that either foster or inhibit a movement organization's ability to serve as a vehicle for communicating the concerns of its members from their lifeworld into the public sphere. We can examine the relationships between discourse, strategy, and organizational form to see what conditions foster a grassroots organization and what conditions foster an Astroturf organization. Finally, the means of resource mobilization can provide an indication as to whether external channeling of the organization by outside funding is occurring.

5

The Growth and Institutionalization of the Environmental Movement

This new commitment [to fighting pollution] has many features of a fad: a rapid swell of enthusiasm (most of the ecology action groups are less than 6 months old), fanned by the mass media. And the commitment is rather shallow.
—Amitai Etzioni, editorial, *Science,* April 1970

It is entirely possible that when the history of the twentieth century is finally written, the single most important social movement of the period will be judged to be environmentalism.
—Robert Nisbet (1982: 101)

The ability of the environmental movement to re-create the public sphere is based on its overall influence in society, which derives from its ability to mobilize a wide variety of symbolic and material resources over a sustained period.

It is generally assumed that environmentalism is a significant social movement in the United States. However, there has been no real analysis of the material resources mobilized by the movement to support that contention.

In this chapter I develop an empirical description of the population of the nationwide environmental organizations.

Overall Size of the Movement

The size of the U.S. environmental movement has never been measured accurately. There is little or no agreement on what constitutes a "movement" organization, or on where the boundaries between the environmental movement and related movements (for example, animal rights) should be drawn.

The U.S. Internal Revenue Service's coding system provides one way to measure the size of the environmental movement. Each organization that has an annual income of more than $25,000 per year is required to file an information tax return. This form contains a block where the organization can designate itself in terms of categories that describe various purposes and activities. The following analysis is based on the use of IRS data records.

Number of Organizations

The first analysis performed was to determine the total number of environmental organizations within the IRS file. Nine IRS categories were selected as identifying environmental organizations. The frequency of these codes were examined. The results of this analysis are shown in table 5.1.

Approximately 10,000 environmental organizations are registered with the IRS as tax-exempt organizations. Many small groups do not bother to register with the IRS, so this figure should be considered the minimum estimate of the number of these groups.

Economic Assets

The total annual expenditures of environmental organizations were calculated from numbers found in the IRS database of nonprofit organizations,

Table 5.1
Number of U.S. environmental organizations, 1995 (based on IRS tax-exempt organization data file dated "12/5/94").

IRS category	Code	Number	Percentage
Preservation of natural resources	350	6,164	59.5
Combating or preventing pollution	351	742	7.2
Land acquisition for preservation	352	381	3.7
Soil or water conservation	353	230	2.2
Preservation of scenic beauty	354	382	3.7
Wildlife sanctuary or refuge	355	410	4.0
Other conservation activities	379	1,674	16.2
Ecology or conservation advocacy	529	354	3.4
Population control advocacy	541	28	0.3
Total		10,365	

which lists each organization's annual income and its assets. The results are shown in table 5.2.

The total financial resources controlled by U.S. environmental organizations listed in the IRS database is approximately $2.7 billion in annual income. The total assets of these organizations amount to $5.8 billion. The distribution of these resources varies significantly among these organizations. Table 5.3 groups the organizations on the basis of income.

The vast majority of environmental organizations have no reportable income. Less than 3 percent of the organizations have annual incomes exceeding $1 million.

Table 5.2
Economic resources of U.S. environmental organizations, 1995.

IRS category	Number	Number with income	Annual income	Assets on hand
Preservation of natural resources	6,164	1,732	$1,992,968,841	$3,812,186,219
Combating or preventing pollution	742	224	$227,992,216	$1,057,778,486
Land acquisition for preservation	380	136	$60,724,791	$162,224,260
Soil or water conservation	230	78	$18,425,426	$29,629,357
Preservation of scenic beauty	38	114	$32,145,179	$111,640,637
Wildlife sanctuary or refuge	410	153	$57,292,840	$107,472,858
Other conservation activity	1,674	511	$243,831,040	$403,647,016
Ecology or conservation advocacy	354	120	$49,545,774	$87,352,924
Population control advocacy	28	17	$412,239,150	$7,328,718
Total	10,365	3,085	$2,695,165,257	$5,779,260,475

Table 5.3
Distribution of economic resources of U.S. environmental organizations, 1995.

Annual income	Number of organizations
None	7,280
$1–$24,999	173
$25,000–$49,999	675
$50,000–$99,999	633
$100,000–$249,999	707
$250,000–$499,999	389
$500,000–$999,999	228
$1,000,000–$9,999,999	245
$10,000,000–$49,999,999	27
Over $50,000,000	8
Total	10,365

Membership

The calculation of the total membership of environmental organizations was developed using estimates of membership based on organizational budget developed elsewhere (Brulle 1995a: 165–169). Membership and staff data is generally not available for organizations with less than $100,000 in annual income, so this analysis is limited to the larger organizations. The estimates developed from these procedures should be seen as probably underestimating membership and staff levels. The average number of members for each budget level was multiplied by the number of organizations in this category. The results of these calculations are shown in table 5.4.

As table 5.4 shows, the total membership of environmental organizations is approximately 41 million. The majority of these members are in the 473 organizations with annual incomes between $500,000 and $9.9 million. To verify this estimated number of members, it was compared with the results of the General Social Surveys (NORC 1994) conducted in 1993, in which 156 out of the sample of 1600 respondents (9.75 percent) said that they were members of environmental organizations. Multiplying the adult population[1] of the United States in 1995 (194,015,000) by 9.75 percent yields an estimate of 18,916,000 members of environmental organizations.

Table 5.4
Membership by budget level, U.S. environmental organizations, 1995.

Estimated budget	Number of organizations	Average number of members	Total number of members
$50 million +	8	675,000	5,400,000
$10 million–$49.9 million	27	427,000	11,529,000
$1 million–$9.9 million	245	64,000	15,680,000
$500,000–$999,999	228	7,000	1,596,000
$100,000–$499,999	1,096	6,000	6,576,000
Total	1,604		40,781,000

One possible explanation for the difference between that number and the calculated estimate of 41 million is multiple memberships. The General Social Surveys do not differentiate between single and multiple memberships. This would contribute to a lower estimate of the total membership in environmental groups by this survey. Using the GSS as the lower bound and the total calculated in table 5.4 as the upper bound, the overall membership of environmental organizations can be said to range between 19 million and 41 million.

Staff Levels

Staff-level estimates were developed in a similar manner. The staff level for each budget level was multiplied by the number of organizations in each budget. This yielded the total number of persons employed as staff by environmental organizations (table 5.5).

Comparison to Other Types of Organizations

In order to gain a perspective on the relative size and importance of the U.S. environmental movement, we must compare it to other social movements and interest groups. The comparison presented here is based on the number and the types of interest groups in the United States, total contributions to several types of nonprofit organizations, household contributions to various causes, levels of volunteer labor, and the environmental movement's influence in a specific policy area.

Table 5.5
Total staff by budget level of organization, U.S. environmental organizations, 1995.

Estimated budget	Number of organizations	Average staff	Total staff
$50 million +	8	800	6,400
$10 million–$49.9 million	27	143	3,861
$1 million–$9.9 million	245	28	6,860
$500,000–$999,999	228	11	2,508
$100,000–$499,999	1,096	8	8,768
Total	1,604		28,397

Other Nongovernmental Organizations

Using the IRS data file described above, three additional categories of organizations were gathered. The results of this analysis are shown in table 5.6. That table shows that the environmental movement is larger and financially better off than either the civil rights movement or the peace movement. However, in comparison to business promotion organizations, the environmental movement has only half as many organizations and only one-fourth as much annual income. This illustrates the predominance of organized business interests over ecological concerns in political debates.

Federal Interest Groups

In his examination of federal interest groups, Walker (1991) developed a typology of these groups. He then measured their size, using a 1985 survey of 863 national interest groups. Walker divided the interest groups into four sectors (ibid.: 10–12, 59). The first, designated the Profit Sector, is composed of groups formed to pursue shared economic interests. This includes trade associations and industry groups. It comprises 37.8 percent of interest groups at the national level. The second sector, with 5.8 percent of the total, comprises organizations that pursue both shared economic interests and broad goals of social change. Labeled the Mixed Sector, it is composed primarily of trade unions. The third sector, which Walker labels the Nonprofit Sector, comprises 32.5 percent of the total and consists of nonpartisan organizations, such as nonprofit health clinics, that provide social services. The fourth sector, labeled the Citizen Sector, comprises 23.9

Table 5.6
Comparison of nonprofit organizations, by category.

	Number	Number with income	Annual income	Assets on hand
Environmental	10,365	3,085	$2,695,165,257	$5,779,260,475
Civil rights	3,169	920	888,474,265	777,556,784
Peace	5,846	2,040	242,985,151	155,274,466
Business promotion	24,984	11,398	8,582,931,005	7,874,257,134

percent of the total and consists of organizations formed as a consequence of political mobilization. As this distribution shows, the organizations formed for mutual economic benefit are the most prevalent type of interest group at the federal level. The Citizen Sector, which encompasses all social-movement organizations, including environmental organizations, is significantly smaller than any of the other sectors.

Walker also examined the issue focus of the Citizen Sector groups. Each group was asked to indicate its level of interest in ten policy areas. One of these ten areas was "Energy/Natural Resources." Just over half (105) of the citizens' groups surveyed indicated that they were either "somewhat" or "very" interested in energy and natural resources. This places these interests seventh out of the ten categories. This question was also asked of all the for-profit organizations. The response of these groups showed a 62.3 percent level of interest in environmental issues (ibid.: 71). Multiplying the number of for-profit organizations by this percentage (326×0.623) yields 203 profit organizations with some interest in environmental issues at the federal level. This indicates a 2-to-1 ratio of profit organizations over citizens' groups in environmental issues at the federal level. This compares almost exactly with the results of the relative preponderance of business promotion organizations to environmental organizations (table 5.6).

Charitable Contributions

A third way to measure the relative importance of the environmental movement is to compare contributions from the general public to environmental groups and to other charitable organizations.

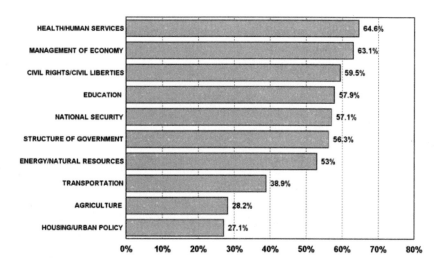

Figure 5.1
Percentages of citizens' groups "somewhat" or "very" interested in various policy areas, based on a 1985 survey of 198 groups (Walker 1991).

Table 5.7 shows the total levels of contributions from all sources in 1995. Environmental and wildlife causes ranked seventh out of eight types of nonprofit organizations. Out of nearly $144 billion contributed to non-profits in 1995, environmental organizations received $4 billion. This amounts to less than 3 percent of all charitable contributions. Religious giving is predominant over all other forms of charitable giving, with well over half of all charitable income going to religious organizations.

Charitable Contributions by Households
The fourth area examined was households' economic support for environmental organizations (table 5.8). The "total contributions" column was calculated by multiplying the number of households in the United States (98,990,000) by the percentage of households contributing to develop a number of total households contributing, then multiplying the result by the average contribution.

As table 5.8 shows, religious giving accounts for more resources than all the other categories combined. Environmental nonprofit organizations receive contributions from 11.5 percent of all households—11,383,850 households contributed an average of $106 apiece, for a total of just over

Table 5.7
Total contributions to nonprofit organizations, 1995 (based on table 616 in 1997 *Statistical Abstract of the United States*).

	Total contributions (billions)	Percentage of total
Religion	$63.5	44.13
Education	17.9	12.44
Health	12.6	8.76
Human service	11.7	8.13
Arts, culture, humanities	10.0	6.95
Public, social benefit	7.1	4.93
Environment, wildlife	4.0	2.78
International	2.1	1.46
All others	15.1	10.49
Total	$143.9	

Table 5.8
Household contributions to nonprofit organizations, 1995 (based on "charitable contributions by household and household statistics" tables in 1997 *Statistical Abstract of the United States*).

	Percentage of households contributing	Average contribution per household	Total contributions (billions)
Religion	48.0	$868	$41.2
Health	27.3	214	5.8
Human services	25.1	271	6.7
Youth development	20.9	137	2.8
Education	20.3	318	6.4
Environment, wildlife	11.5	106	1.2
Public, social benefit	10.3	122	1.2
Arts, culture, humanities	9.4	216	2.0
Recreation (adults)	7.0	161	1.1
International	6.1	283	1.7
Private community foundations	6.1	181	1.1
Total			$71.0

$1.2 billion. This places environmentalism in a tie for seventh overall, behind religion, health, human services, education, youth development, and the arts. However, when the average contribution per household is examined, environmentalism is ranked last, with an average contribution per household of $106. Per household, religious giving exceeds giving to environmental organizations by a factor of more than 8.

Volunteer Work
Almost half of adult Americans report doing volunteer work. In 1995, 48.8 percent of the population worked an average of 4.2 hours a week for no monetary pay to help others. The distribution of the adult population doing this work is shown in table 5.9. Environment-oriented activities rank eighth overall.

Influence in the Federal Risk Community
The final area of comparison of the relative strength of the environmental movement is based on an analysis of the federal risk community that appears in *The Risk Professionals* (Dietz and Rycroft 1987). The risk com-

Table 5.9
Participation in volunteer work, 1995.

Type of activity	Percentage of U.S. population involved
Religion	25.8
Education	17.5
Youth development	15.4
Health	13.2
Human services	12.7
World-related organizations	7.9
Recreation (adults)	7.3
Environment	7.1
Public-societal benefit	6.7
Arts, culture, humanities	6.2
Political organizations	3.8
Private and community foundations	2.7
International, foreign	1.6

munity is defined as professionals who work in the policy area of environmental health and safety risk at the national level. In a series of interviews with 228 risk professionals conducted in 1984, Dietz and Rycroft found that 89 (39 percent) were employed by the federal government. Subtracting that number from the total leaves 139 professionals employed by groups outside the government. The distribution is shown in table 5.10.

On the basis of the data on page 20 of Dietz and Rycroft 1987, it appears that the majority of consulting and law firms perform work for corporations. Hence, these totals are included as one category to represent the business community. Based on this distribution, the staff levels of industry outnumber those from environmental organizations by a ratio of 3 to 1. This lends additional support to Walker's (1991) conclusion that business interests have a significant advantage over environmental organizations in terms of number of organizations. As the analysis by Dietz and Rycroft shows, this advantage also carries over into the number of staff in an environmental policy area.

In a further examination of the risk community, Dietz and Rycroft (1987: 81–86) measured "prominence" in communication flows in this community (figure 5.2). Prominence is a measure of the number of times members of groups contact members of another type of group. It is an empirical measurement of the importance of a type of organization in the network of communication. The figure shows that corporations and law and consulting firms have higher prominence than environmental groups.

Another measure developed by Dietz and Rycroft (ibid.: 98) was the overall influence of the type of group in the process of policy development. To measure influence in the risk community, each individual was asked to

Table 5.10
Employment of risk professionals (adapted from Dietz and Rycroft 1987: 20, table 2).

	Number employed	Percentage of total
Corporations, consultants	72	52
Environmental groups	27	19
Think tanks, universities	21	15
Other	19	14
Total	139	

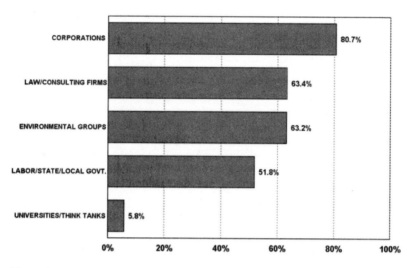

Figure 5.2
Prominence of interest groups in communication flows in the risk policy community, based on Dietz and Rycroft 1987: 81–86. (Prominence: percentage of times members of risk community report contacting employees of various types of organizations.)

rate the influence of other groups. The results were used to measure the overall influence of groups throughout the community. Industry ranked first, with a 64.9 percent of the respondents rating this group as "very" influential. Environmental groups were rating as "very" influential by 55.3 percent of the respondents. Labor was rated "very" influential by only 13.5 percent of the respondents.

The difference between environmental organizations and industry groups is more pronounced when influence within the federal government is measured. The ratings of the federal government members of the risk community of the relative influence of industry, environmental, and labor groups (table 5.11) show the high level of perceived influence of industry by government employees. This perceived influence exceeds that of environmental groups and labor groups in all areas of government. The overall difference between industry's and environmental groups' perceived influence is just over 9 percent. However, EPA employees rate industry as 20 percent more influential than environmental groups. In other executive

Table 5.11
Influence of interest groups as perceived by federal risk professionals (adapted from Dietz and Rycroft 1987: 98, table 8).

Government group	Interest group		
	Industry	Environmental	Labor
EPA	68.6	48.6	2.9
Other executive agencies	82.8	48.3	11.1
Congress	80.0	44.0	25.0
Overall rating	64.9	55.3	13.5

agencies and in Congress, this difference increases to approximately 40 percent.

Comparative Size of the Environmental Movement

From the above five comparisons, the comparative size of the environmental movement can be gauged in three ways. First, the environmental movement is larger than either the civil rights movement or the peace movement in terms of number of organizations, income, and assets. Thus, it is probably one of the largest social movements in the United States. However, industrial interest groups have a consistent advantage over environmental groups, both in number of organizations and in number of persons employed by these organizations. In addition, the perceived influence of industry is seen as higher than that of environmental organizations in all the branches of government that deal with issues of environmental risk. Second, in comparison to other nonprofit organizations, the environmental movement has a significant presence in the policy-making process. The poll taken by Walker (1991) placed environmental groups seventh in terms of number of citizens' group interested in this policy area. Third, the level of contributions places environmental organizations at a very low level in comparison to almost all other types of nonprofit organizations. Religious organizations dwarf environmental organizations, collecting eight times as much per household and perhaps 20 or 30 times the level of annual donations. Thus, while the environmental movement may be a significant social movement, it has significantly less resources than either business or nonmovement nonprofit organizations.

Conclusions

The assumption that environmentalism is a significant social movement is generally borne out by the analysis in this chapter. The following conclusions can be made regarding environmental organizations in the United States in 1995.

The total number of U.S. environmental organizations is estimated to be approximately 10,000. These 10,000 organizations have the between 19 million and 41 million members, approximately 28,000 staff employees, total annual income of $2.6 billion, and assets of $5.8 billion.

The environmental movement is a major social movement in the United States—perhaps the largest one. However, it has significantly less resources than either religious or business organizations. Whether or not the environmental movement has sufficient resources to re-create the public sphere is an open question.

6

Manifest Destiny and the Development of North America

Our manifest destiny is to overspread the continent allotted by Providence for the free development of our yearly multiplying millions.

—John Louis O'Sullivan, *New York Morning News*, December 27, 1845

America's resources were put here for the enjoyment and use of people, now and in the future, and should not be denied to the people by elitist groups.

—James Watt, *Nation's Business*, September 1981: 36–41

From 1620 to the middle of the nineteenth century, the dominant and virtually unchallenged discourse that guided Americans' relationship with the natural environment was the discourse of Manifest Destiny (Norton 1991: 76–77; Marcus 1992: 46–47; Short 1989: ix; Gedicks 1993: xi). The current U.S. environmental movement originated in protests against this discourse. Thus, to understand the development of this movement requires an appraisal of the dominant discourse that preceded its emergence.

The Discourse of Manifest Destiny

Manifest Destiny is a summary term for the dominant discourse with regard to nature in the United States. This discourse provides both an economic and a moral rationale for exploiting the natural environment (Norton 1991: 76–77). Key components of this discourse are the following:

- Nature has no intrinsic value.
- Human welfare is based on development of the natural environment.
- The natural environment is unproductive and valueless without development.
- Human labor transforms the natural environment into useful commodities.

- There are abundant natural resources available for human use.
- Humans have a right to use the natural environment to meet their needs.

In addition to providing the rationale for the development of the North American continent, Manifest Destiny has continued to serve as the discourse of several waves of countermovements opposed to the goals and aims of the various U.S. environmental organizations. The social movements based on the discourse of Manifest Destiny can thus be seen as countermovements.

Countermovements are "networks of individuals and organizations that share many of the same objects of concern as the social movements that they oppose" (Meyer and Staggenborg 1996: 1632). They "make competing claims on the state on matters of policy and politics and vie for attention from the mass media and the broader public" (ibid.). These types of movements arise in response to social movements for change. In effect, the advocates of change create their own opposition. By advocating an alternative discourse and challenging established interests, they create conditions for mobilization of their opposition. Countermovements originate as the change movement starts to show signs of success by influencing public policy, and threatening established interests (ibid.: 1635–1640). The elites of these interests then respond to this threat by fostering countermovements to protect their interests by opposing or containing the challenging social movements (Pichardo 1995: 23). As Gale (1986: 207) notes, these countermovements "typically represent economic interests directly challenged by the emergent social movement." The countermovement organizations that emerge take the form of elite driven efforts to mobilize economically impacted populations, or populations that share similar interests or ideologies (ibid.: 207).

The discourse of a countermovement generally takes the form of appeals for the restoration of the "established myths of society" (Lo 1982: 119). Within U.S. environmental politics, the debate has generally centered on challenges to and advocates for the dominant discourse of Manifest Destiny. As this analysis will show, the struggles over which discourse defines the correct relationship of humanity and nature have taken place over 100 years and continue today.

The Development of Manifest Destiny

The development of Manifest Destiny has deep roots in American culture. The Pilgrims saw North America as "a hideous and desolate wilderness, full of wild beasts and wild men" (Short 1989: 1). The Pilgrims, and the vast number of settlers that followed, set out to settle the North American continent. Nash (1967: 24) described this attitude as a moral play in which "civilizing the New World meant enlightening darkness, ordering chaos, and changing evil into good" and "in the morality play of westward expansion, wilderness was the villain, and the pioneer, as hero, relished its destruction."

The Native Americans were also seen as part of this uncivilized nature. Fueled by the additional discourse of racism, this led toward genocide. An example of this type of belief is found on page i of *American Pioneer History* (Mason 1896): "Civilization is a war—a war of light with darkness; of truth with falsehood; of the illuminated intellect and the rectified heart with the barbarism of ignorance and the animalism of the savage."

Exploitation of the natural environment and removal of the Native Americans were seen as both a moral duty and an inherent right of a "civilized" people. While this attitude was probably an unquestioned reality for many, it was finally defined in an 1845 article by John Louis O'Sullivan: "Our manifest destiny is to overspread the continent allotted by Providence for the free development of our yearly multiplying millions." (Bartlett 1968: 676) Coupled with free-market ideology, this made exploitation of the natural environment into a duty. Moral and economic gain were defined through exploitation of the natural environment. In 1846, capturing the spirit of the time, the publicist and land speculator William Gilpin wrote: "The untransacted destiny of the American people is to subdue the continent—to rush over this vast field to the Pacific Ocean . . . —to establish a new order in human affairs. . . . Divine task!" (Gilpin 1846/1973: 124)

Some of the states' seals contain strong representations of this attitude toward the natural environment. A state seal can be seen as a graphic representation of the aspirations of the population, or at least of the political elites at the time the seal is adopted. The ideal of subduing nature for commercial purposes comes through clearly in Indiana's seal: an industrious pioneer is shown chopping down the ancient forest while a buffalo jumps over a fallen log. The seal of Kansas shows a farmer industriously plowing up the

Figure 6.1
The state seals of Indiana and Kansas.

prairie. Just beyond the farmer's field, a covered wagon moves onto the prairie, chasing buffalo before it. In the background, a side-wheel steamer pours smoke into the atmosphere as it carries its cargo down the river.

These attitudes of mind have persisted in our culture. One example of these beliefs was reported in the 1930s. In a study conducted by the Congressional Great Plains Committee, the numerous causes of the Dust Bowl were examined. In one part of the report, the role of individual attitudes was examined as one of the root causes of this ecological degradation:

Rehabilitation of a great region in which it has been discovered that economic activities are not properly adjusted to basic and controlling physical conditions, is not merely a problem of encouraging better farm practices and desirable engineering works, and revision of such institutions as ownership and tenure. It is also one of revision of some of the less obvious, deep-seated attitudes of mind. (U.S. Great Plains Committee 1936: 63)

This report then went on to describe the instantiation of attitudes of Manifest Destiny in the Great Plains, and to call for their revision. These attitudes of mind were identified as having eleven key components (ibid.: 63–67)[1]:

1. Man conquers nature.
2. Natural resources are inexhaustible.
3. Habitual practices are best.
4. What is good for the individual is good for everybody.

5. An owner may do with his property as he likes.
6. Expanding markets will continue indefinitely.
7. Free competition coordinates industry and agriculture.
8. Property values will increase indefinitely.
9. Tenancy is a stepping stone to ownership.
10. The factory farm is generally desirable.
11. The individual must make his own adjustments.

This discourse has continued to animate collective behavior regarding the environment throughout the twentieth century (Short 1989: ix). Prominent manifestations of this discourse are the movements against the expansion of government environmental-protection efforts in the western United States. The history of western land struggles shows that there have been a number of these conflicts. These protests follow a distinct pattern. First, in response to demands from the environmental movement, the federal government intervenes in the economy and issues regulations over the control of either federal lands or the production process. These interventions trigger rebellions and protests, primarily in the western states. This cycle of economic development, protests by environmental organizations, development of federal regulation, and countermovement led by economic elites adversely impacted by the federal action has occurred several times in U.S. history. Seven of these conflicts are discussed in this section.

The Struggle over the Creation of National Forests
The first struggle between environmental movements and countermovements occurred with the initial development of the forest reserve system. In 1891 and 1892, President Harrison created the first six national forests. Although there were some editorials protesting these actions, there was no organized opposition to them. On February 22, 1897, President Grover Cleveland created thirteen new national forests. This action increased the acreage from the original 13 million acres set aside by Harrison to 39 million acres.

Massive protests erupted in the west. On February 26, a rally of more than 30,000 people protesting the forest reserves occurred at Deadwood, South Dakota (Robbins 1962: 316). A bill was introduced in Congress to restore these reservations to the public domain (Dana and Fairfax 1980: 60). Echoing the sentiments of western states, Senator Carter of Montana accused the president of "contemptuous disregard" for the people's interest (Richardson 1962: 1). Governor Richards of Wyoming disdainfully denounced the creation of forest reserves, as they had been "created by college professors and landscape gardeners" (Richardson 1962: 8). Senator Wilson of Washington asked "Why should we be everlastingly and eternally harassed and annoyed and bedeviled by these scientific gentlemen from Harvard College?" (Robbins 1962: 323)

The struggle against the development of national forests continued and intensified between 1907 and 1916 (Maughan and Nilson 1993: 2). On March 6, 1907, President Theodore Roosevelt, on the advice of Chief Forester Gifford Pinchot, designated 159 forest preserves and added 148 million acres to the national forests.

Further protests occurred, and the first steps toward collective political action began to emerge. The key event of the protest was the Public Lands Convention held in Denver in June of 1907. This was a protest meeting of more than 861 delegates against the federal presence in western lands (Richardson 1962: 36–40). Arguments were made for the first time to turn the federal land over to the states; thus initiating a long debate over the proper disposition of federal lands that continues today.[2]

The Public Lands Convention and other protests in the west over government land policy were one of the reasons why, in May of 1908, President Roosevelt called a Conference of Governors on Conservation. One of the unstated goals of this conference was to provide a political outlet for the brewing dissatisfaction in the western states with the development of forest reserves. This conference succeeded in dampening some of the protests.

Part of this campaign was the formation of the National Public Domain League. This league was formed in June of 1909 through the efforts of the governor of Colorado, John Shafroth, and the organizers of the Public Lands Convention of 1907. Its major efforts were to publish a series of brochures to attack the idea of conservation, and its chief leader, Gifford Pinchot.

One example appears in Bulletin No. 8, Conservation Alarms, published on August 16, 1909. On its cover was a comic drawing of Pinchot conjuring up the various goblins of famine in the imagination of the western states. Inside the front cover was a long poetic parody of Little Orphan Annie's song "The Gobble-uns 'at gits you ef you don't watch out." The first stanza reads as follows:

Little Prophet Pinchot
By Eli Snograss

Little Prophet Pinchot is seein' things, they say,
He's seein' 'em at night-times an' he's seein' 'em by day
He's seein' 'em 'ats mighty short, an's seein' 'em 'ats small,
He's seein' what we've all-us seen, but never seen at all.
An' ef we takes some papers up-some magazine or trac'
The fust we knows we're dippy, an' our souls are on the rack.
A-readin' of the scare-tales 'at Gifford tells about,
An' the Famin-uns 'gits you
        Ef you
            Don't
                Watch
                    Out!

Figure 6.2
The goblins of Pinchot (from cover of Conservation Alarms, Bulletin 8, Public Domain League, August 16, 1909).

In the autumn of 1909, another similar group was formed, the Western Conservation League. Its purpose was to "to obtain justice for the West by securing state control of resources" (Richardson 1962: 90). Like its predecessor, the Public Domain League, it was primarily engaged in publicity affairs. Neither organization had an active membership, nor substantial financing, and did not grow into significant collective action organizations (Graf 1990: 130–131).

In 1911 a second Public Lands Convention was held. However, just over 100 delegates attended. By 1914, with the beginning of World War I, the development of conservation policy was no longer a major political issue, and the first countermovement subsided (Richardson 1962: 155).

The Stanfield Rebellion

Another series of western protests over federal land policy occurred during the period 1925–1934. Led by Senator Robert Stanfield of Oregon, this rebellion was a protest against the imposition of grazing fees in the national forests (Cawley 1993). In 1925 and 1926, Senator Stanfield held a series of widely publicized hearings about the adverse impact of grazing fees. These hearings focused national attention on western lands policy and fueled a series of bills and hearings in Congress. A law was introduced in Congress in 1932 to return federal lands to the states. Although these efforts were ultimately unsuccessful, they set the stage for inclusion of cattlemen in the management of federal grazing policy.

Federal grazing policy was firmly established with the passage of the Taylor Grazing Act in 1934 (Maughan and Nilson 1993). This law established the development and regulation of grazing districts on federal lands. Its purpose was to use scientific management to "stop injury to the public grazing lands" (Clepper 1966: 140). However, it also incorporated western cattlemen into the management practices of public lands in the form of the National Advisory Board, which advised the Grazing Service on how best to manage its lands. The Taylor Grazing Act was further amended in 1939 to establish advisory boards of cattlemen for each grazing district (U.S. Department of the Interior 1959: 22). The stated purpose of the National Advisory board council was to "further the interest of users of the public domain under the jurisdiction of the Grazing Service" (Foss 1960: 121). In effect, this greatly improved the power of ranchers to allocate and manage federal lands for grazing.

The McCarran Protests

While the passage of the Taylor Grazing Act of 1934 settled most of the disputes over western land use, there were still a number of cattlemen, especially in Nevada, who continued to protest the use of grazing fees. This protest was led by Senator Pat McCarran of Nevada. This protest gathered further momentum when the Grazing Service proposed a tripling of grazing fees. During the period 1941–1946, McCarran held a series of hearing to highlight the adverse impact of grazing fees on stockholders. In addition, in 1946, Senator Allen Robertson of Wyoming reintroduced the bill originally introduced in 1932 to transfer all federal lands to the states. These protests were resolved in late 1940s with the demise of the Grazing Service, and the creation of the Bureau of Land Management, with a much-reduced staff (Graf 1990: 166). This weakened the ability of the federal government to enforce the required grazing fees.

In a series of hearings before the Subcommittee on Public Lands, Congressman Barrett of Wyoming also tried, unsuccessfully, to stir up momentum to abolish the grazing service. In one hearing, held in Ely, Nevada, on October 4, 1947, one of the members of the committee compared environmental controls over grazing to communism: "The power of the government to regulate grazing seems more nearly modeled on the Russian way of life, and though we are opposing Russian autocracy, in government regulations of the range we are building that very same system. To protect the ranges, the forest and the watersheds is communism." (DeVoto 1948: 53)

The end result of the protests of the stockmen was to greatly strengthened the role of cattlemen in the governance of public land. Their ability to manage this land almost entirely for grazing remained unchallenged until the mid 1960s.

The Campaign against *Silent Spring*

In the autumn of 1962, Rachel Carson's landmark book *Silent Spring* was published. A concerted effort on the part of industry to argue against the conclusions of the book ensued. A number of editorials and publications appeared denouncing the book. A major leader in this publicity campaign was Robert White Stevens of the American Cyanamid Corporation. Beginning in December of 1962, Stevens gave a series of speeches to dispute the accuracy of Carson's work. In 1963, in a televised speech,[3] Stevens said: "The major claims in Miss Rachel Carson's book, *Silent Spring,* are gross distortions of the actual facts, completely unsupported by scientific experimental evidence and general practical experience in the field." Stevens also displayed faith in the ability of science and man to control nature, and he disputed the idea of the balance of nature and the need to live within its bounds: "The crux, the fulcrum over which the argument chiefly rests, is that Miss Carson maintains that the balance of nature is a major force in the survival of man. Whereas the modern chemist, the modern biologist, the modern scientist believes that man is steadily controlling nature."

The campaign backfired. The publicity fired interest in *Silent Spring,* and sales of the book soared. This conflict was resolved in late 1963 when *Use of Pesticides,* a report prepared by the president's science advisors, was published. This report supported virtually all of Carson's analysis, and it silenced public criticism of her book.

The Sagebrush Rebellion

In the 1960s and the 1970s, the laws governing the use of land in the west shifted in favor of the environmental movement. The Wilderness Act permanently locked up parts of national forests. The Endangered Species Act placed limits on land use, even by private property owners. The Bureau of Land Management, responding to new legislative initiatives, shifted its management of federal lands to consider multiple uses of public lands other than grazing. The National Advisory Board also shifted to include significantly different constituencies who represented competing uses of the public lands other than grazing (Cawley 1993: 75). These changes were perceived as an expansion of environmental interests at the cost of traditional western economic interests. The measure of this increased power of the environmental community was based on "the proliferation of regulations during the 1970s that exhibited a distinct bias in favor of environmental protection values" (ibid.: 161).

In response to this situation, another western protest emerged in the period 1979–1983. Its initial identity was fixed by the term "Sagebrush Rebellion" (Shabecoff 1993: 164). Basically, the agenda was a reincarnation of the old land issues raised in the previous rebellions (Graf 1990: 228). The Sagebrush Rebellion began in July of 1979, when the Nevada state legislature approved a resolution that demanded the return of 49 million acres of federal land to state control (Short 1989: 13). This was an attempt to roll back the environmental laws developed in the 1970s, primarily the Federal Land Policy Management Act of 1976.

Groups representing western economic interests, primarily ranchers and miners, made a major effort to return control of federal lands to local economic interests. This appeal was based on three arguments: that individual states would be better land managers, that economic growth in the western states was stifled by restrictive federal land management, and that states had the right to control the land within their borders (Short 1989: 15). The ultimate strategy behind this movement was to privatize federal lands by transferring them to the states, and then to private hands (Cawley 1993: 103).

In the late 1970s, several groups came into being to carry out this initiative. According to a 1994 handout, the Center for the Defense of Free Enterprise was founded on July 4, 1976, out of concern "about the multitude of restrictions that were being imposed on America's free enterprise system by big government and the lack of understanding of this problem by the American people." The Mountain States Legal Defense Fund was founded in 1977, with the notorious anti-environmentalist James Watt as its president and chief legal officer. This organization has been characterized as a business-supported anti-environmental-law firm (Graf 1990: 243).

An organized collective countermovement appeared at a conference of the League for the Advancement of States' Equal Rights (LASER) in Salt Lake City in January of 1981. LASER was "engaged in creating a broad base of public support in favor of divesting the federal government of the public domain" (LASER Proceedings 1981: 1). Cawley (1993: 161) has characterized this countermovement at this time as follows: ". . . the Sagebrush Rebellion represented a protest against the growing influence of the environmental community. . . . The question confronting Sagebrush Rebels, therefore, was how to curtail the environmental community's influence and therefore stem the tide of regulations." That question was answered with the election of Ronald Reagan to the presidency. Reagan made James Watt his Secretary of the Interior. With that appointment, the stage was set for a dramatic rollback of environmental regulations.

During Ronald Reagan's first term, there were a number of legislative proposals to weaken existing environmental laws. In addition, the budgets of government programs to protect the environment were severely reduced. These threats to environmental laws created great concern within the environmental movement, and there was a surge of membership in environmental groups. This increased level of mobilization enabled the movement to block many of the initiatives proposed by the Reagan administration. Essentially, the actions of the Sagebrush Rebellion and the Reagan administration was to produce a stalemate in the area of environmental policy. According to Cawley (1993: 161): "Lacking sufficient support to eliminate objectionable environmental policies, the administration was able to slow the environmental movement's momentum. Conversely, the environmental community could not eliminate the threat posed by the administration, but it did block implementation of most of Reagan's environmental agenda." The momentum of the Sagebrush Rebellion faded at the end of Reagan's first term, when two of its leading proponents in government, James Watt and Ann Gorsuch, were forced to resign after significant allegations of illegal activities (Szasz 1994: 116–130).

The Wise-Use Movement

The issues raised by the Sagebrush Rebellion were given new impetus and expanded toward the end of the late 1980s under a more comprehensive movement know as the wise-use movement (Cawley 1993; Knox 1990; O'Callaghan 1992; Stapleton 1993). Helvarg (1994: 9) saw the wise-use movement as a "counterrevolutionary movement, defining itself in response to the environmental revolution of the past thirty years." The intellectual roots of this movement combine the idea of Manifest Destiny, states' rights, and property rights to call for the reversal of environmental restrictions. Expanding on the advocacy of the free market for dealing with environmental problems developed by Milton Friedman (1962; see also Cawley 1993: 137–141 and Callahan and Myer 1992: 6), the wise-use movement argues that market mechanisms can best manage all natural resources.

In his 1985 book *Takings: Private Property and the Power of Eminent Domain*, Richard Epstein expanded this idea to encompass virtually all government actions. Epstein (ibid.: 95) made the case that the Fifth Amendment to the U.S. Constitution requires that the government pay compensation whenever its actions adversely affect the property rights of an individual:

"All regulations, all taxes, and all modifications of liability rules are takings of private property prima facie compensatable by the state." This would effectively prohibit the enactment of any environmental regulations by requiring large expenditures for economic compensation for individuals effected by these actions.

The wise-use movement took organization form at the Multiple Use Strategy Conference, organized by the Center for the Defense of Free Enterprise and held in Reno in August 1988 (Cawley 1993: 166; Helvarg 1994: 76; Stapleton 1992 1993; Knox 1990; O'Callaghan 1992).[4] *The Wise Use Agenda* (Gottlieb 1989), which emerged from the Reno conference, listed five broad policy goals:

1. Adopt policies that will promote both the use of the environment and wisdom in that use: Adequate natural resources for our social and economic well-being developed by means that protect the natural environment.
2. Identify and encourage wise use technologies that work in productive harmony with nature.
3. Promote public awareness of outstanding examples of man working in harmony with nature.
4. Discourage extremism in both resource use and protection. National policy should officially recognize that human economic development can be carried out by means that enhance the natural environment.
5. Administer wise use policy with equal sensitivity toward natural values and human values.

The document then lists the top 25 goals of the movement. One goal is "all public lands including wilderness and national parks shall be open to mineral and energy production" (Gottlieb 1989: 6). Another is amending the Endangered Species Act to allow for differential treatment of "relic species in decline before the appearance of man" (ibid.: 12).

One of the earliest public demonstrations by this movement, held on May 21, 1989, at Oregon's capital building, in Salem, protested an injunction against logging in the habitat of the spotted owl (Ramos 1995: 88). This was followed by a number of other demonstrations organized by this movement. The characteristics of these demonstrations was one of careful orchestration. Tokar (1995: 153) has noted in studying these demonstrations that "in the early 1990s, the highest-profile activities of anti-environmental groups were carefully staged demonstrations of loggers and millworkers in northern California and the Pacific Northwest in opposition to the protection of old growth forests and the habitat of the endangered north-ern spotted owl."

The wise-use movement continued to expand in the 1990s. It is estimated that the movement had 200 groups in 1988 and more than 1500 groups nationwide in 1995.

The movement's three main components are property-rights advocates, who oppose federal regulations on private land ownership and use; county-supremacy groups, primarily located in the west, that seek to replace fed-eral with county control over public lands; and general-purpose national groups, who seek to limit federal environmental laws of all sorts (Brick and Cawley 1996: 7).

The wise-use movement is based on a coalition of "loggers, ranchers, miners, farmers, fishermen, oil and gas interests, real estate developers, and off-road vehicle enthusiasts" (Brick and Cawley 1996: 7). Its overall goals are "to create and mold disaffection over environmental regulations, big government, and the media that can win respectability for centrist argu-ments seeking to "protect jobs, private property and the environment by finding a balance between human and environmental needs" (Helvarg 1994: 9). In addition, it pushes a more radical agenda based on the rhetoric of free-market environmentalism, privatization of public lands, and the removal of environmental regulations from industry.

To realize its agenda, this group carries out a wide variety of actions, including education and lobbying. It also has developed the strategy of suing its opponents for libel in public speeches. Known as SLAAPS (Strategic Litigation Action Against Public Speech), this strategy has taken a toll on the ability of environmental activists to openly voice their opinions (Canan and Pring 1988; Helvarg 1994; Grumbine 1994).

In March of 1998, the wise-use movement distributed an article titled Environmental Effects of Increased Atmospheric Carbon Dioxide. Closely resembling a reprint from the *Proceedings of the National Academy of Sciences,* this article concluded that predictions of global warming were in error, and increased CO_2 levels greatly benefit plants and animals. In fact, this article was a nonrefereed publication, funded by the George C. Marshall Institute, a wise-use think tank opposed to CO_2 emission curbs. In the opinion of many scientists in the field, this was a deliberate attempt to mislead the public regarding the true state of global warming.

In a 1993 assessment, the social scientists Ralph Maughn and Douglas Nilson called the wise-use movement "a desperate effort to defend the hegemony of the cultural and economic values of the agricultural and extractive industries of the rural west" that "differs from past such movements in its level of desperation and in a first effort to win allies in other parts of the region and nation" (Maughn and Nilson 1993: 1).

Wise-Use Organizations

Organizations based on the discourse of wise use define their objective in language such as "preserving our natural resources for the public instead of from the public," "protecting the American way of life," and "preserving private property rights." They use phrases such as "scare tactics of the environmentalists" and "free market allocation of values." Their publications focus is on how the use of public lands provides economic and recreational benefits for different groups without adversely harming the land. Their analysis of environmental issues is generally slanted to dispute the supposed harm of different environmental problems, especially the issue of global warming. Their magazines generally contain advertisements for four-wheel-drive and off-road vehicles.

To examine the characteristics of organizations informed by this discourse, I compiled and examined the by-laws and income-tax returns of thirteen wise-use groups.[5] On the basis of the sample, the following general comments can be made.

While the founding dates range from 1953 (Keep America Beautiful) to 1992 (Alliance for America), most wise-use groups are relatively young. The majority were founded in the late 1980s and the early 1990s, and the

Table 6.1
Sources of income of wise-use groups.

	Percentage of income
Public contributions	90.9
Government grants	1.9
Program services	3.1
Membership dues	2.6
Other	1.5

average[6] founding year is 1983. The organizations are, for the most part, small or moderate in size and income. The average group has 56,000 members and a full time staff of 15. The annual average income of these groups is $1,984,111, and the average assets are $278,311. As table 6.1 shows, the majority of the income comes from direct public contributions. This includes contributions from both foundations and corporations. Internal Revenue Service rules do not require the sources of contributions to be specified, so it is very difficult if not impossible to ascertain the proportions of these contributions that come from individuals, from corporations, and from foundations. However, the extent of corporate influence can be demonstrated. The Environmental Working Group of Washington, D.C., has traced many of the sources of the funding of these groups to corporations. An analysis of the EWG's database shows that the 13 groups in the sample derived their income from an average of 133 different corporations. The EWG lists more than 4000 different organizations that fund wise-use groups. Among the top 50 are ten major oil corporations, eight chemical corporations, six logging companies, four automobile manufacturers, and three mining corporations. Seventy percent of these groups use parliamentarian[7] and other existing channels of dispute resolution as their main means of pursuing their goals, 20 percent use mainly educational means, and 10 percent use mainly protest tactics. Sixty percent of the groups are oligarchic in organizational structure, 10 percent are representative, 20 percent have limited democracy, and 10 percent are fully democratic.

Social Learning and the Wise-Use Movement

The discourse of Manifest Destiny continues to be a key component of the worldview of the wise-use groups.

The model of countermovements as elite-driven attempts to rein in an expanding challenging movement is an excellent perspective from which to understand the wise-use movement and its organizations. There is good evidence that a significant proportion of the funding of these organizations comes from corporations with economic stakes in the policies the movement focuses on. The preponderance of oligarchic structures also reinforces the existence of centralized control of these groups.

The discourse of Manifest Destiny has animated and informed many of the most ecologically destructive practices in the United States. It provides economic, nationalistic, and religious justifications for ecologically destructive practices. It is opposed by virtually all environmental organizations.

However, this discourse can sensitize our society of the human and community impacts of our actions to preserve the natural environment. The development of actions to combat ecological degradation will have to be based on a dialogue in which all the interested and affected parties participate. Although many persons may oppose the style, organizational makeup, and tactics of the wise-use movement, this component of U.S. society represents the economic corporations and communities that are affected by actions to protect the environment. Thus, it has a right to have its concerns voiced and attended to in the formulation of actions to protect the natural environment. What is needed is the development of organizations that are not dominated and controlled by economic corporations but rather are authentic and legitimate representatives of the affected human communities.

7

The Early Development of the Environmental Movement

The development of the environmental movement began before the American Civil War, when a number of concerns over the destruction of the natural environment were voiced. These concerns were later manifested in movements for the protection of game, the conservation of forests, and the preservation of wilderness. By 1900, all three of these areas had been institutionalized into distinct movement organizations, government bodies, various protected areas, and a body of laws and regulations.

In the period from 1920 to the mid 1960s, these activities came to be summarized as the Conservation Movement. This term tended to collapse the significant differences among the communities of the environmental movement (Fox 1981: 130), much as Environmentalism stands for all the movement's communities today.

Wildlife Management

Wild beasts and birds are by right not the property merely of the people of today, but the property of the unborn generations, whose belongings we have no right to squander.
—Theodore Roosevelt, 1915[1]

Hunting is the harvesting of a man-made crop, which would soon cease to exist if somebody somewhere had not, intentionally or unintentionally, come to nature's aid in its production.
—Aldo Leopold, 1932[2]

Concern over the natural environment first appeared in the United States over the issue of hunting. Reiger (1975: 11–24) convincingly argues that the contribution of American sportsmen to both the development of the national parks system and the preservation of wildlife is generally over-looked by the field of environmental history. Other authors (Leopold 1933: 13–21; Trefethen 1964: 9–13; Gilbert and Dodds 1987: 7–11) have presented ample evidence regarding the role of sportsmen as an important component in the development of wildlife conservation, the national forests, and the development of the national parks system. Hence, the origins of some important aspects of the conservation and preservation movements can be traced to the efforts of American sportsmen.

The discourse of Wildlife Management defines both a unique viewpoint and the practices of a distinct community of organizations. It defines a distinct community centered on wildlife conservation issues, which includes nongovernmental organizations such as the American Wildlife Institute, Ducks Unlimited, and the National Wildlife Federation; government agencies such as the U.S. Biological Survey; and academic programs in wildlife management or wildlife conservation (Tober 1989). The following are key components of this discourse:

• The scientific management of ecosystems can ensure stable populations of wildlife.
• This wildlife population can be seen as a crop from which excess populations can be sustainably harvested in accordance with the ecological limitations of a given area.
• This excess wildlife population thus can be utilized for human recreation in sport hunting.
• The scientific management of ecosystems can ensure stable populations of wildlife.

The Early Development of Game Protection

Hunting practices have their basis in tribal taboos (Leopold 1932: 5). Virtually all previous societies have had some restrictions or rules on the harvesting of game (Gilbert and Dodds 1987: 1–3). An early codification of this appears in Deuteronomy 22:6: "If a bird's nest chance to be before thee in the way, in any tree or on the ground, with young ones or eggs, and the dam sitting upon the young, or upon the eggs, thou shalt not take the dam with the young."

The American tradition and ethics toward hunting originated in the British sporting tradition. This tradition is based on the perceived value of hunting as a noble occupation and as training for warfare (Manning 1993: 57). Thus, hunting was the proper domain of the nobility. John Lyly commented in his 1592 play *Midas* that hunting was "for kings, not peasants." To allow the nobility unhindered access to hunting, hunting preserves called "forests" were established. One of the earliest instances occurred in 1062, when a "charter of the forest" was granted by Canute the Dane. This charter described a forest as "a circuit of woody grounds and pastures, known in its bounds as privileged for the peaceable being and abiding of wild beasts and fowls for forese, chase, and warren, to be under the king's protection for his princely delight" (Leopold 1932: 10). A "sportsman's ethic" developed informally in England (MacKenzie 1988); it took an aristocratic form, and it included proper social etiquette for hunting. As described by Reiger (1975: 26), this meant that hunting had to be performed in accordance with a strict social code:

Whether fly-fishing for trout in an English stream or grouse shooting on a Scottish moor, an aristocrat took his sport seriously. To be fully accepted by his peers, he had to have a knowledge of the quarry and its habitat; a familiarity with the rods, guns, or dogs necessary to its pursuit; a skill to cast or shoot with precision and coolness that often takes years to acquire; and most of all, a "social sense" of the do's and don'ts involved.

Game Protection

This sportsman's ethic was brought to America and took root in the upper class. Sportsmen, as they preferred to be called, formulated the initial concerns over the plight and decimation of American wildlife in the 1840s (Reiger 1975; Bean 1978). This concern centered on the term *game protection*, which originated in the British sportsman's tradition. This term conveyed a concept of controlling access to a scare resource by limitation of hunting. The point of this limitation was to avoid the decimation of particular game species (Trefethen 1964: 1–20; Gilbert, and Dodds 1987: 6–11). This perspective was dominant throughout the latter half of the nineteenth century and the first 30 years of the twentieth century.

The first American book on the ethics of hunting, *The Sportsman's Companion*, was published anonymously in 1783 (Phillips and Hill 1930: xiii). As early as 1830, the American author T. Doughty described the characteristics of a true sportsman: "A true Sportsman always respects the rules and seasons for shooting, and most heartily despises the man who destroys the unfledged brood, or the protectors which Nature has provided for them." (Doughty 1830: 7)

One of the most influential writers in this tradition was the British aristocrat H. W. Herbert, whose pen name in the United States was Frank Forester. Herbert, born in London in 1807, was educated at Cambridge University. He emigrated to the United States in 1831 (Clepper 1971: 159–160). His articles began appearing in magazines in the late 1830s, and his first book, *Warwick Woodlands*, was published in 1845. Herbert, an avid sportsman, strongly advocated the British culture of sportsmanship as the appropriate code of conduct for hunting. This led to the appearance of "gentleman hunters" who promoted the "virtues" of hunting in a quest for status (Schmitt 1969: 13; Norton 1991: 24; Reiger 1986).

Forester also criticized existing hunting practices. For example, in his *Introductory Observations to Field Sports of the United States* he observes the ignorance of hunting and comments that "the rarest and choicest species are slaughtered inconsiderately . . . at such times and in such manners, as are rapidly causing them to disappear and become extinct " (Forester 1873: 12). This concern resulted in a "Code of Sportsmen," which Hummel (1994: 13) summarized as follows:

1. The sportsman should practice proper etiquette in the field—not crowding other hunters, not claiming more than a fair share of shooting opportunities, not taking more game than can be consumed by the sportsman.
2. Do not pursue any game species to the point of extinction.
3. Develop skill with the weapons of the hunt so that they can deliver a humane coup-de-grace to the prey species.
4. Acquire extensive knowledge of the prey species and its habitat.
5. Allow the prey a fair opportunity to escape by use of its natural defense/flight mechanisms. Meet the prey in its own environment and master it by "fair means."
6. Possess an appreciation for the sport in order to be its model representative to the non-sporting public and to pass it on to future generations.

The writings of these early authors were not limited to sportsmanship. As Reiger (1975: 26) points out, the sport hunters "did more than outline the basics of sporting etiquette; they also lamented the commercial destruction of wildlife and habitat."

The Development of Local and State Groups
The concern for the plight of American wildlife developed into a number of state and local "Sportsmen's Clubs" to foster the development and

practice of game protection. The first group founded on this was the New York Sportsmen's Association (*Forest and Stream* no. 33, 1889: 450). This organization was founded in 1844. Among its first members was H. W. Herbert (Grinnell and Sheldon 1925: 222). The 1889 version of their constitution (NY State Association for the Protection of Fish and Game 1884: 31) stated the reasons for its formation:

This Association is formed for the purpose of securing proper legislation for the protection of wild birds, fish, and animals throughout the State during the season at which it is improper to pursue them, for the vigorous enforcement of such laws as shall be enacted, and for the promotion of kindly intercourse and generous emulation among sportsmen.

The symbol of the New York Sportsmen's Association "displays a woodcock in flight, with the legend *Non nobis solum*—not for ourselves alone—thus suggesting, perhaps for the first time, the ideal now accepted by all sportsmen that the gunners of one generation are, in fact, trustees of the game, to hold and use it for their time and to hand it down to those who are to follow" (Grinnell and Sheldon 1925: 223–224).

By 1878, Hallock's American Club List (Hallock 1878: 61–68) listed 317 Sportsmen's Clubs. These clubs had a vast influence at the state and local levels in the development and implementation of hunting laws and regulations. The first game preserve was established 1871 by the Blooming Grove Park Association in Pike County, Pennsylvania (Reiger 1975: 57). A rapid expanse of state and local game laws also followed. In 1850, there was "comparatively little game legislation" (Palmer 1912: 10). Only 19 states

Figure 7.1
Emblem of the New York Association for the Protection of Game.

had some limitations on hunting. By 1880, all 48 states and territories had game-protection laws. Between 1910 and 1910, 324 state game laws were passed (ibid.: 10–12).

The Emergence of a National Game-Protection Movement
The initial federal efforts in game protection centered on the development of the U.S. Fish Commission. At the urging of sportsmen and market fishermen, this agency was established in 1871 to investigate and protect the Connecticut River fishery (Reiger 1975: 53).

A national game-protection movement was slow to develop. The next national organization to undertake efforts and lobbying in this area was the Society for the Prevention of Cruelty to Animals. A campaign protesting the slaughter of the buffalo, run by the SPCA's president, Henry Bergh, led to the introduction of a bill by Congressman Richard McCormick of Arizona on March 13, 1871 (H.R. No. 157). After being passed by both houses of Congress, it was subjected to a pocket veto by President Ulysses S. Grant (Grinnell and Sheldon 1925: 215). In examining the reasons for the failure of this bill to be signed into law, Garretson (1934: 37) stated that "it was commonly understood that the U.S. Government was interested in the rapid extinction of the buffalo; the basis on which the independent existence of the Plains Indian depended."

This early effort was followed by the founding of the magazine *Forest and Stream* by George Grinnell and Henry Hallock in 1873. This magazine served as a national voice for the concerns of sportsmen. This magazine, under the editorship of Henry Hallock, was an outspoken voice for game protection. In editorials, Hallock "reminded the American public week after week of what rapid industrialization was doing to the natural environment" (Reiger 1975: 31). According to Reiger (ibid.), Hallock preached along the following lines:

The blind worship of PROGRESS, the nation's secular religion, was at the root of the problem. Men must be taught to "enjoy the present Earth and the present life, so that there shall be less necessity to look for the promised creation of 'a new heaven and a new Earth.'"

In one of his first actions, Grinnell used the magazine *Forest and Stream* to initiate a campaign against hunting in Yellowstone. It also voiced concern over forestry and habitat loss and advocated conservation and management of forests. For example, in an 1873 article, Hallock wrote: ". . . the culti-vation of forests . . . and the selling of timber and surplus game of all kinds will . . . compensate in some degree for the frightful waste which is annu-ally devastating our forests and exterminating our game" (Hallock 1873, quoted in Reiger 1975: 57).

The need for a national organization to fight for game-protection laws was also recognized by George Grinnell, and in the February 11 1886 issue of *Forest and Stream* Grinnell announced the founding of the Audubon Society (Grinnell and Sheldon 1925: 230–231; Reiger 1975: 66–69). The purpose of this organization was "the formation of an association for the protection of wild birds and their eggs" (Reiger 1975: 68). "Its member-ship," Grinnell continued, "is to be free to every one who is willing to lend a helping hand in forwarding the objectives for which it is formed. These objects shall be to prevent, so far as possible, (1) the killing of any wild birds not used for food, (2) the destruction of nests or eggs of any wild bird, and (3) the wearing of feathers as ornaments or trimming for dress." (ibid.) Because of his workload, Grinnell was unable to run Audubon Society and *Forest and Stream*. The initial Audubon Society was disbanded in 1889. However, in the following years, several state Audubon associa-tions were founded. The National Association of Audubon Societies—now the National Audubon Society—was reconstituted in 1905 with Grinnell as national director (ibid.: 69).

In 1887, the Boone and Crockett Club was founded by Theodore Roosevelt. The purposes of this organization, as quoted on pp. 4–5 of Grinnell 1910, were

(1) To promote manly sport with the rifle. (2) To promote travel and exploration in the wild and unknown, or but partially known, portions of the country. (3) To work for the preservation of the large game of this country, and so far as possible to further legislation for that purpose, and to assist in enforcing the existing laws. (4) To promote inquiry into and to record observations on the habits and natural history of the various wild animals. (5) To bring about among the members inter-change of opinions and ideas on hunting, travel and exploration; on the various kinds of hunting rifles; on the haunts of game animals, etc.

The early members of the Boone and Crockett Club were highly influential individuals, including President Theodore Roosevelt, several congressmen and senators (including Henry Cabot Lodge and John Lacy), the painter Albert Bierstadt, the industrialists J. P. Morgan Jr. and John J. Pierpont, and several writers and editors (Grinnell 1910: 65–71).

The Boone and Crockett Club was one of the leading advocates for the development of a national forest reserve system. Grinnell first used the notion of conservation to advocate protection of forests and watersheds in the Adirondack Mountains. On January 17, 1884, he urged "protection and conservation, now, prompt, adequate." "This," he continued, "is what the Adirondack forests demand, not restoration years hence, after the damage from unregulated lumbering shall have been wrought and ruin has followed." (Reiger 1975: 85)

At the national level, the Boone and Crockett Club was instrumental in the development of the National Forests. John W. Noble, a club member and Secretary of Interior under President William Henry Harrison, was highly involved in the development and eventual passage of the Forest Reserve Act of 1891 and the designation of the first forest reserves by President Harrison (Reneau and Reneau 1993: 7). In 1902 the Boone and Crockett Club lobbied for successful passage of a law protecting Alaskan game (Grinnell 1910: 37–41), and in 1910 it led the successful effort to establish Glacier Park (Grinnell 1910: 48–49). And the Boone and Crockett Club was instrumental in the development of 14 new groups, including the American Bison Society, the New York Zoological Society, the Audubon Society, More Game Birds for America (renamed Ducks Unlimited in 1937), Wildlife Management Institute, and the National Wildlife Federation (Grinnell 1910: 31–35; Trefethen 1961: 120–122; Boone and Crockett Club 1995: 9; Ward and McCabe 1988: 97). In 1885, upon the urging of the American Ornithologists' Union, Congress authorized the development of the U.S. Biological Survey. In 1894, Congress passed the Yellowstone Park Protection Act, which prohibited hunting in the park (Trefethen 1964: 10–11)

Although there were numerous state laws regulating hunting, there were no federal laws in this area until May 25, 1900. On that date, the U.S. government first became directly involved in wildlife issues with the passage of the "Lacey Act" (Statutes of the United States, 56th Congress: 187). The purpose of this act was strictly utilitarian: "to ensure an adequate supply of hunting and Insectivorous birds: to aid in the restoration of such birds in those parts of the United States adapted thereto where the same have become scarce or extinct, and also to regulate the introduction of American or foreign birds or animals in localities where they have not heretofore existed."

Professionalization of game management followed quickly. In 1902 the International Association of Game, Fish, and Conservation Commissioners was founded (Trefethen 1961: 121). In 1911 the American Game Association was founded to develop scientific game-management programs. This organization, funded partially by gun and ammunition dealers to ensure the growth of recreational hunting, was transformed into the Wildlife Management Institute in 1935 (Trefethen 1961: 122). In 1903 the Pelican Island Reservation, the first national wildlife refuge, was established (U.S. Department of the Interior 1959: 11).

The most notable conservation group to be founded in the 1920s was the Izaak Walton League, begun in 1922 by a group of fishers and hunters. This organization, primarily based in the Midwest, focused its efforts on the traditional concerns of forests and forestry and on water pollution and the loss of wetlands (the latter because of their adverse impacts on fishing) (Fox 1981: 159–160).

In 1929 the Audubon Society came under criticism as an organization "with a large income and a cash surplus which owing to entangled alliances performs its work with inertia, incompetency and procrastination" (Fox 1981: 174). This critique spawned a movement to reform the society. To carry out this work, the suffragist Rosalie Edge created an organization known as The Emergency Conservation Committee. Edge based her critique of the Audubon Society's indifferent defense of other living creatures on "man's fatuous egotism before the natural world" (Fox 1981: 176). Her purpose was to "break open the tunnel vision imposed by a man centered view of the universe" (ibid.: 181). She was successful in bringing major changes to the Audubon Society.

The Shift to Wildlife Management

At the beginning of the 1930s, the discourse of game protection was redefined to encompass scientific management of ecosystems to ensure stable populations of wildlife. One of the central figures in this redefinition was Aldo Leopold, a member of the Boone and Crockett Society. Leopold's landmark book *Game Management*, published in 1932, marks the development of the current discourse of Wildlife Management, which he defined as "making land produce sustained annual crops of wildgame for recreational use" (Leopold 1932: 3) or as "the art of cropping land for game" (ibid.: ix).

Leopold (1932: 16–17) argued that game protection's core idea—that restriction of hunting could "string out" the remnants of the virgin supply and make them last longer to perpetuate rather than to improve or create hunting—was too limited in scope. Game protection, he asserted, ignored the results that scientific management of game could achieve in developing a perpetual supply of game animals for hunting. "In short," Leopold wrote (1933: 210), "the attempt to control hunting has suffered . . . from the persistence of the concept that all hunting is the division of nature's bounty. We must replace this concept with a new one: that hunting is the harvesting of a man-made crop, which would soon cease to exist if somebody somewhere had not, intentionally or unintentionally, come to nature's aid in its production." The basis for the management of wildlife was the application of the concepts of ecology to the management of wild areas. Leopold argued that a given habitat could support only a fixed amount of wildlife, and that the additional population of wildlife above this level should be available to sport hunters.

New academic courses of instruction in wildlife management were developing. The first cooperative wildlife education program was developed in Iowa in 1932 by J. N. "Ding" Darling. This program was expanded in 1935 to encompass cooperative Wildlife Research Units at nine land-grant colleges. Aldo Leopold became the first professor of game management at the University of Wisconsin.

This shift in discourse coincided with changes in the practices of wildlife management (Gilbert and Dodds 1987: 8–10, 17; Trefethen 1964 16–17). To implement the new approach of active management of game animals, a number of new wildlife-protection laws were passed in the 1930s. One of the key acts was the 1934 Duck Stamp Act. In purpose this act was nearly identical to the Lacey Act, enacted 34 years earlier; however, it provided for the collection of fees to support game refuges. All persons hunting migratory birds were required to purchase a stamp. The funds from the sale of Duck Stamps would be used "for the location, ascertainment, acquisition, administration, maintenance, and development of suitable areas for inviolate migratory-bird sanctuaries" (Statutes of the United States, 73rd Congress, March 16, 1934: 451).

In addition, there were changes within existing game-protection organizations, and several new organizations focused on wildlife conservation were established. The American Game Protective and Propagation

Association, founded in 1911, became defunct in 1935 (Haskell 1937). The annual American Game Conference was terminated in 1935.

In February of 1936, the first North American Wildlife Conference was convened by President Franklin Roosevelt. Roosevelt defined the purpose of this conference as "to bring together individuals, organizations, and agencies interested in the restoration and conservation of wildlife resources" (Proceedings of the North American Wildlife and Natural Resources Conference 1936: III). The conference was attended by nearly 2000 delegates (Allen 1987: 29). There were presentations by more than 100 representatives of federal, state, and local government agencies, non-governmental organizations, academics, politicians, and corporations. Twelve sessions on special topics of concern were held. Among the topics discussed were Farmer-Sportsman Cooperatives, Fish Management, Forests and Forest Wildlife, Pollution, and the Problem of Vanishing Species (Proceedings of the North American Wildlife and Natural Resources Conference 1936: IV–XI). The outgrowth of this conference was "the ascendance of wildlife management as the strategy for the preservation of wildlife" (Gilbert and Dodds 1987: 8). At this conference, the constitution of the National Wildlife Federation was drafted and the Wildlife Society was formed (Gilbert and Dodds 1987: 9).

The first North American Wildlife Conference was followed by a flurry of activity. Later in 1936, the Biological Survey and the Bureau of Fisheries were combined to form the U.S. Fish and Wildlife Service. In 1937 the Pittman-Robertson Federal Aid in Wildlife Restoration Act was passed. This law provided funding for the development of wildlife sanctuaries.

The foundations for the current laws and organizations in the area of wildlife management were thus firmly established in the late 1930s, and they have undergone little significant change since that time. The practices of game management have continued to be developed through the development of the discipline of wildlife management. However, the focus of this discursive frame, and the organizations based on it, have remained essentially unchanged.

Current Wildlife-Management Organizations

Organizations based on the discourse of Wildlife Management define their objective as conserving or rationally developing wildlife resources to provide for human recreation. They use phrases such as "maximizing the supply of game" and "conserve our wildlife resources." Their magazines are usually highly polished and illustrated with the game species that each organization seeks to conserve.

To examine the characteristics of organizations informed by this discourse, I compiled a sample of seven different wildlife-management groups.[3]

There are two clusters in the founding dates of these organizations: one in the mid 1930s and one in the early 1980s. The mean founding year is 1947. These organizations are, for the most part, extremely large. The National Wildlife Federation and Ducks Unlimited are very large in terms of both income and membership. These two groups skew the averages for the rest of the groups. The average group has 861,000 members and a full-time staff of 172. The annual average income is $25,967,443. The average assets of one of these groups exceed $7 million. As table 7.1 shows, the majority of the income comes from public contributions and membership services.

As a strategy for influencing policy, 83 percent of the groups use primarily education. The other 17 percent use primarily parliamentarian or other existing channels of dispute resolution. Thus, these groups exclusively use conventional means to influence the policy process.

In organizational structure, 50 percent of the sample groups have representative government and another 17 percent are limited democracies. The remaining 33 percent have an oligarchic organizational structure. The prevalence of representative forms of government in these groups is probably due to their historical pattern of development. Many of these groups were formed by merging several state organizations into a national con-

Table 7.1
Sources of income of wildlife-management groups.

	Percentage of income
Public contributions	32.6
Foundation grants	3.5
Government grants	2.3
Program services	7.0
Membership dues	21.1
Other	33.5

federation. In creating this group, each chapter was given some form of individual representation in the governance of the organization.

Social Learning and Wildlife Management

Wildlife Management is a limited and specialized discourse. It sees nature as a resource for human utilization. Based on utilitarian and anthropocentric considerations, it values nature only for its value to humans in recreational activities. This severely limits the range of issues and concerns addressed by the organizations that use this discourse. Their narrow issue focus does not lead them to address wider issues of ecological degradation.

The appeal of these organizations to a wider audience is limited. People who do not hunt or fish are not likely to seek membership in these organizations.

These organizations engage in traditional political activities and do not advocate wide-ranging social change. Because the bulk of their economic support comes from members, these organizations are highly unlikely to be significantly influenced by external funding organizations.

Despite their narrow focus, these organizations provide an instructive example of large, member-funded, participatory environmental groups that authentically represent their members' concerns.

Conservation

The first principle of conservation is development, the use of the natural resources now existing on this continent for the benefit of the people who live here now. . . . In the second place conservation stands for the prevention of waste. . . . The third principle is that the natural resources must be developed and preserved for the benefit of the many, and not merely for the profit of a few. . . . Conservation means the greatest good to the greatest number for the longest time.
—Gifford Pinchot, 1910

Sustainable development [means] to ensure that it meets the needs of the present without compromising the ability for future generations to meet their own needs.
—World Council on Environment and Development (WCED 1987)

Perhaps the most influential early discursive frames in the U.S. environmental movement was developed by the conservation movement. It defines a utilitarian and technical/managerial perspective regarding nature (Oelschlaeger 1991: 201–202, 286–289). From the viewpoint of conservationism, nature is a resource to be used by society to meet human needs. This forms the basis for collective action to ensure that natural resources are used by applying the criteria of rationality and efficiency to achieve the maximum utility to society. The following are among the key components of this perspective:

• Physical and biological nature is nothing more than a collection of parts that function like a machine.
• Humans need to use the natural resources provided by nature to maintain society.
• Nature can be managed by humans through the application of technical knowledge used by competent professionals.
• The proper management philosophy for natural resources is to realize the greatest good for the greatest number of people over the longest period of time.

The Development of the Idea of Conservation

The roots of the idea of conservation can be traced to sixteenth-century Europe. In the writings of this time, man, as the highest being in creation, had both a special responsibility and a privilege. God's design was that man was responsible to maintain nature as the steward of God's creation. In addition, man was to act as God's partner in creation by "improving the Earth through his inventiveness, for his enjoyment and use, and for the greater glory of God" (Glacken 1956: 64). These beliefs, when combined with the technical literature on natural resources, created a philosophy and a practice of the scientific management of natural resources that came to be known as conservation.

Both the practical and the ethical dimensions for the conservation of nature took form in the seventeenth century (Harre et al. 1999: 12–15). As Grove (1990, 1992) notes, the adverse ecological impacts of colonization of tropical islands brought the need for government action to conserve nature into consciousness and political action. Tropical islands were seen as existing edenic paradises. However, as these islands were developed, the effects of capitalism and colonial rule on the natural environment became apparent. This gave rise to the perceived need for ecological understanding of human effects and the need for land control. Some of the first actions to preserve nature were taken by colonial authorities. The French authorities took action in 1769 to preserve significant parts of the natural environment in Mauritius. This was followed by British action in the West Indies in 1791 (Grove 1992: 44–45).

The work of the German geographer Alexander von Humboldt consolidated the literature on the relationship between humans and the natural environment (Humboldt 1849, 1850a,b). Humboldt was instrumental in developing an integrated vision based on a number of biological, geological and economic studies. As a result, his work helped to "promulgate a new ecological concept of the relation between people and the natural world: that of the fundamental interrelation of humankind and other forces in the cosmos" (Grove 1992: 45).

The discipline of forestry was most developed in Germany, where it had taken the form of a science of natural resource management. German forestry was based on three key assumptions summarized by Bennett (1968: 5–20; Dana and Fairfax 1980: 52):

1. As the product of a limited land base, wood is scarce; 2. Wood is a necessity for which there will always be a demand, and because of unchanging social patterns and consumption patterns, the demand for wood is stable; 3. Because wood is scarce and demand is stable, the appropriate standards for forest management are certain: limit the consumption of wood to the growth potential of the forest.

Based on these assumptions, the German foresters developed the idea of managing a forest so that it produces a sustained yield of wood, equal to the growth of the forest (Dana and Fairfax 1980: 52). This idea of the management of natural resources to provide a sustained yield was generalized to cover the management of all natural resources, including soil, water, and wildlife.

The initial development of conservation in the United States is generally seen to have originated in 1894 with the publication of George Perkins Marsh's book *Man and Nature* (Nash 1990: 40; McCormick 1989: 11; Oelschlaeger 1991: 106, 282). Building on the work of the "new" geographers, especially Humboldt, Marsh argued that humanity was "a destabilizing environmental force whose impacts portended an uncertain future" (Oelschlaeger 1991: 107). Therefore, humanity must develop a stewardship of natural resources. Marsh viewed the threats to the environment as being caused by three factors. As summarized by Taylor (1992: 16–17), these three factors were as follows: "First, there is too little understanding of natural processes and the relationships that keep nature in balance. . . .

Second, the attitude toward the Earth (as a limitless source of resources) must change . . . , and third, that one of the institutions most guilty of flagrant disregard for nature is the modern business corporation."[4] Marsh's book was a warning to "his countrymen—by comparing environments of old with those of newly settle regions—of the dangers of disturbing the balance of nature" (Glacken 1956: 63–66) "The lands of the New World," Glacken continues, "would require careful husbanding if they were to remain habitable for future generations."

Another major contributor to this idea was John Wesley Powell, whose *Report on the Lands of the Arid Region* was one of the most significant documents in American conservation history (Dana and Fairfax 1980: 39). Powell described the conditions of the western landscape and argued for a utilization plan based on adaptation to the natural conditions. In the preface to the first edition, Powell (1878: 8) writes: "To a great extent, the redemption of all these lands will require extensive and comprehensive plans. . . . It was my purpose not only to consider the character of the lands themselves, but also the engineering problems involved in their redemption."

From these early beginnings, the idea of conservation took root. One factor in this was the need for some sort of control over the reckless exploitation of natural resources. Another was the increasing concern in the technical, managerial, and professional elites about the availability of natural resources (Nash 1967: 265–276). The early conservation movement took the form of a top-down effort to develop a program of resource conservation to ensure the continuation of economic development (Oelschlaeger 1991: 286–288; McCormick 1989 p. 13; Hays 1985: 202). This was linked to the development of a professional, technically managed economy. According to Hays (1959: 2–3), "conservation, above all, was a scientific movement, and its role in history arises from the implications of science and technology in modern society. . . . Its essence was rational planning to promote efficient development and use of all natural resources." The guiding idea of this effort was "the use of foresight and restraint in the exploitation of the physical sources of wealth as necessary for the perpetuity of civilization, and the welfare of present and future generations" (ibid.: 123).

Beginnings of the Conservation Movement

The development of the conservation movement gained national attention in 1873 when Franklin Hough, a physician from upstate New York, read a paper at the 22nd meeting of the American Association for the Advancement of Science. The paper, On the Duty of Governments in the Preservation of Forests, was an eloquent argument for the development and application of scientific forestry in the United States. Hough concluded with a call for the AAAS to advocate conservation to U.S. political leaders:

These questions are not limited to a particular state, but in the interest the Nation generally; and I would venture to suggest that this Association might properly take measures for bringing to the notice of our several State Governments, and Congress with respect to the territories, the subject of protection to forests, and their cultivation, regulation and encouragement; and that it appoint a special committee to memorialize these several legislative bodies upon this subject, and to urge its importance. A measure of public utility thus commended to their notice by this Association would doubtless receive respectful attention. Its reasons would be brought up for discussion, and the probabilities of the future, drawn from the history of the past, might be represented before the public in their true light.

In 1874, following Hough's recommendation, the AAAS wrote a letter to Congress and to President Grant. This letter, "after pointing out the great public injury likely to result from the rapid exhaustion of the country's forests, . . . requested the creation of a commission of forestry which would report directly to the president" (Dana and Fairfax 1980: 42). In 1876, Congress authorized the Secretary of Agriculture to "study the present and future demand for timber and other forest products, the probable supply for future wants, the means best adapted to their preservation and renewal, and to file a report on his inquiry with Congress" (ibid.: 50). The person selected to develop this study was none other than Franklin Hough.

Hough produced a three-volume *Report upon Forestry* (Hough 1878, 1880, 1882) that detailed the current state of U.S. forests, the economics of forestry, and the need to prevent forest fires. Throughout the report, he advocated the development of forest reserves to enable scientific management of the forests (Dana and Fairfax 1980: 50–51). In 1886 Congress acted on Hough's reports, creating a Division of Forestry in the Department of Agriculture. A Prussian-trained forester, Bernhard Fernow, became the first Chief Forester. Fernow advanced an idea he called "timber physics," which later grew into the discipline of forestry (ibid.: 53). He was also an influential lobbyist for the development of forest reserves.

The American Forestry Association, founded in 1875, also advocated forest reserves (Dana and Fairfax 1980: 42–43; Clepper 1966: 9). Now known as American Forests, this organization, at its inception, had the following as one of its major objectives:

The forests of the country should be made to yield the greatest possible benefits to present and future generations, both by producing timber crops and by less direct means. Lumbering is an inseparable factor of the best forest protection and management and should be so conducted as not to destroy the productive capacity of the land. It is believed that saw logs, mine timbers, railroad ties, etc. can be cut without the usual accompanying destruction of the forests. This Association will endeavor to promote all these lumber interests. (*The Forester* 5, January 1899: 1)

This organization's purpose has changed little in the last century. Its current bylaws list its primary objective as "to bring about more efficient conservation and development of forests and other natural resources toward realization of the maximum value of these resources in the economic, industrial, and social progress of the nation" (American Forests Bylaws, as amended November 7, 1992).

The first governmental actions to establish forest preserves was taken in 1885, when the state of New York designated fourteen counties in the Adirondack Mountains as a forest preserve. This was followed at the federal level in 1887 when Edward Bowers of the AFA, in conjunction with Chief Forester Bernard Fernow,[5] developed a bill proposing actions that would have "withdrawn from entry of all public lands valuable in any degree for timber or their forest growth. The reservations would then be administered by a Forestry Bureau, an elaborate agency established to administer an equally elaborate system of limber licenses at an annual cost of $500,000." (Dana and Fairfax 1980: 55) Though this bill was not enacted into law, it set the stage for further federal legislative action.

In 1891 the federal Forest Reserve Act was passed. This was the first federal law establishing authority for forest conservation (Reneau and Reneau: 1993: 7). The key section of this law, section 24, contained a provision that allowed the president to create forest reserves "in any state or Territory having public land bearing forests in any part of the public lands wholly or in part covered with timber or undergrowth, whether of commercial value" (Statutes of the United States, 51st Congress: 1103). This law enabled the forest advocates to urge the creation of a system of national forest reserves. Less than a month after passage of the bill, President Harrison created the Yellowstone Park Forest Reserve. Over the next two years, he proclaimed 14 forest reserves and created a forest reserve system with a total gross area of 13 million acres. On February 22, 1897, another 14 forest reserves were designated by President Grover Cleveland. This added an additional 25.8 million acres to the forest reserve system (Dana and Fairfax 1980: 58–60).

The creation of these reserves created a public uproar among western timber and grazing interests. As a result, a series of amendments to the Forest Reserve Act were proposed. The amendments culminated in the Organic Act of 1897, a rider on an appropriations act for the Department of Agriculture that allowed for logging in national forests subject to the supervision of the U.S. Forest Service. This law established three purposes for forest preserves. To this day it provides the main statutory basis for forest management in the United States (Dana and Fairfax 1980: 63). This law states:

No public forest reservation shall be established, except to: improve and protect the forest within the reservation, or for the purpose of securing favorable conditions of water flows, and to furnish a continuous supply of timber for the use and necessities of citizens of the United States. (Statutes of the United States, 55th Congress: 35)

As the Forest Reserve system was being established, the discipline of forestry was developing. The first book on forestry was published by Franklin Hough in 1882 (Dana and Fairfax 1980: 51). Forestry education in the United States began with lectures at Yale University in 1873. By 1887, courses in forestry were offered at seven universities (ibid.: 43). By 1890, 18 universities offered such courses (Secretary of Agriculture 1890: 224–225). By 1899, there were 49 universities offering courses in 39 states (Secretary of Agriculture 1900: 703–704). In 1900, the Society of American Foresters was founded "to further the cause of forestry in America by fostering a spirit of comradeship among American foresters; by creating opportunities for a free interchange of views upon forestry and allied subjects; and by disseminating a knowledge of the purpose and achievements of forestry" (quoted in Clepper 1966: 45–46).

In addition, forest tracts were established to test and demonstrate the concept of scientific forestry. One of the earliest such tracts originated in

1890 when the Adirondack League Club purchased more than 93,000 acres of ancient forest in upstate New York "with the stated purpose of placing it under forest management" (Secretary of Agriculture 1890: 208). The discussion of the terms of how the forest to be run states that "the cutting must be done with a view to favor reproduction and not in the haphazard way in which the lumberman does it. Here come in the science of forestry" (Secretary of Agriculture 1890: 210). In 1893, Gifford Pinchot established a scientifically managed forest plantation, Biltmore Forest, in North Carolina (Dana and Fairfax 1980: 54).

Roosevelt, Pinchot, and the Institutionalization of Conservation

Throughout the latter half of the eighteenth century, there was a growing movement for the conservation of both natural resources and wildlife. This movement, initiated by sportsmen, was later joined by scientifically trained foresters. It came to its culmination when Theodore Roosevelt was president and Gifford Pinchot was Chief Forester. During that period, even though Roosevelt and Pinchot were "late-comers to conservation" (Reiger 1975: 50), the conservation movement was able to institutionalize the conservation perspective in several government agencies.

The conservation of Pinchot and Roosevelt was part of the overall Progressive movement. This version of conservation emphasized the wise technical administration of natural resources for the enhancement of material life and the support of distributive justice. At its core, Progressive conservation was an attempt to manage natural resources to provide the material conditions for a liberal democracy (Taylor 1992: 1). As Taylor (ibid.: 200) summarizes it, "Pinchot's fight for conservation . . . was actually a fight to protect equality of opportunity for the plain citizen from the privilege of wealth and corporate power." Safeguarding and scientifically managing natural resources were seen as means of providing for the material needs of a democratic society (ibid.: 20). The idea of scientific management of natural resources by dedicated public agencies for the common good created tremendous political support for conservation initiatives (ibid.: 136).

Under the leadership of Pinchot, several political initiatives were taken to start government programs based on this viewpoint. The U.S. Forest Service was founded in 1905. In 1907, Roosevelt designated 159 forest preserves and added 148 million acres to the national forests, almost quadrupling their size. In addition, he withdrew from public use 80 million acres for mining studies, 1.5 million acres for possible water-power sites, and 4.7 million acres of phosphate lands (Van Hise 1921: 2–14), developed the first program of federal irrigation, and initiated the Reclamation Service.

In May of 1908, Roosevelt called a Governors' Conference on conservation. This conference was a major milestone in official recognition of conservation (McCormick 1989: 16; Oelschlaeger: 282–283). Participants included the president, the vice president, nine Supreme Court judges, governors from 34 states, numerous congressmen and senators, representatives of 68 national associations, and hundreds of experts. In his opening speech, Roosevelt defined the purpose of the conference:

This Conference on the conservation of natural resources is in effect a meeting of the representatives of all the people of the United States called to consider the weightiest problem now before the Nation: and the occasion for the meeting lies in the fact that the natural resources of our country are in danger of exhaustion if we permit the old wasteful methods of exploiting them longer to continue. (Blanchard 1909: 3)

Roosevelt went on to call for a policy of the "wise utilization" of natural resources:

The time has come for a change . . . to protect ourselves and our children against the wasteful development of our natural resources. . . . Such a policy will preserve soil, forests, water power as a heritage for the children and the children's children of the men and women of this generation. (ibid.: 10–11)

Over the next three days, a series of meetings were held on the conservation of natural resources, including mineral fuels and ores, soil, forests, sanitation, irrigation, grazing and stock raising, water resource development, and electrical power (ibid.: xiii–xvi).[6]

After the Governors' Conference, there were public proclamations of support for its work from several associations. The May 1908 issue of *Forestry and Irrigation* cites endorsements from numerous state forestry associations, the Eastern States Retail Lumber Dealers' Association, the Los Angeles Chamber of Commerce, the New York Board of Trade and Transportation, the Pomona Grange of Oregon, the Association for the Protection of the Adirondacks, and the Detroit New Century Club. The May 1909 issue of *Annals of the American Academy of Political and Social*

Science was dedicated to the conservation of natural resources. Several women's clubs endorsed and pledged their support for conservation, including the Federation of Women's Clubs, the Daughters of the American Revolution, the General Federation of Women's Clubs, and the Woman's National Rivers and Harbors Congress. In 1911, *American Conservation* (1, no. 4: 346) noted that "the four leading women's organizations of the United States, comprising nearly a million members, and embracing practically all the women's clubs in the country, have taken up conservation." The First Exposition of Conservation and its Builders, held in Knoxville in 1913, celebrated the development of the conservation movement in the United States (Goodman 1914).

Conservation was even celebrated in poetry. In the May 1908 *Journal of Forestry and Irrigation*, several poems appeared, including *The Hill Man's Lament*, *The Cry of the Pines*, *The Brave Old Oak*, *Trees*, and the following:

The Spendthrift
By Robert M. Reese, Washington D.C.

I.
Into my great inheritance
    I came when I was young;
I spent it freely with both hands;
    I mocked, with jeering tougue,
At those who sorrowfully said
    "Beware! The end is near!"
And, drunk with riches, shook my head,
    Regarding not their fear.
II.
My squandered forests, hacked and hewed,
    Are gone; my rivers fail;
My stricken hillsides, stark and nude,
    Stand shivering in the gale.
Down to the sea my teeming soil
    In yellow torrents goes;
The guerdon of the farmer's toil
    With each year lesser grows.
III.
Lord! Of Thy bounty heedless still,
    My store of good I spend;
Thy brimming cup I careless spill,
    Regarding not the end.

My riches melt away like snow
　beneath the April rain;
And though my hand prepareth Woe,
　Yet may I not refrain.
IV.
O stay my sinful hand and lend
　My falt'ring heart Thine aid,
That these my spendthrift days may end
　And at Thy feet be laid
The will to show the past retrieved,
Thy gifts renewed, restored.
That I have spent what I received,
　Thy pardon grant, O Lord!

One of the manifestations of the conservation movement was a movement to improve the quality of rural life. Known as Country Life, this movement sought to improve and modernize rural life while preserving its unique values. Bowers (1974: 4) notes that its adherents hoped "to hold fast to the social and political virtues of the agrarian past while still retaining the material benefits of industrial changes." The Country Life movement was part of the larger Progressive movement, and was led by a group of well-educated urban dwellers. It focused on the improvement of rural education systems, built bicycle paths and parks, and attempted to beautify rural highways. The land-grant colleges were enlisted to provide rural communities with technical skills and knowledge to help them scientifically manage their natural resources and community efforts. Out of this effort, the academic disciplines of rural sociology and agricultural economics came into being. The rural extension services of the Department of Agriculture also developed out of the Country Life movement. This movement ended in the early 1920s as the result of a split between the components that sought to promote farming as a way of life and the growing efforts to improve the economic efficiency of the farm through the further application of industrial farming techniques (Bowers 1974).

Franklin D. Roosevelt and the Advancement of Conservation

The momentum of the conservation movement stalled when Theodore Roosevelt left office, and soon the nation's political agenda was dominated by a world war. In 1928, in one notable exception (foreshadowing the Dust Bowl of 1934), H. Bennett and W. Chapline published "Soil Erosion: A

National Menace," a paper urging action to deal with the growing problem. It was largely ignored.

With the election of Franklin D. Roosevelt as president, the conservation movement again became an active force in federal policy. In 1933, Roosevelt founded the Civilian Conservation Corps to "give young people healthful work and directly conserve important national resources" (Owen 1983: 13). The Tennessee Valley Authority was created as an overall government effort to deal with "all natural resources as a single problem" (ibid.: 15) in one geographic area. In 1934 the Taylor Grazing Act was passed to "stop injury to the public grazing lands" through the use of scientific management (Clepper 1966: 140). Also, several major dams, including Hoover and Grand Coulee Dam were started during the Roosevelt administration.

The major ecological problem of this time was the advent of the Dust Bowl in 1934. Soil conservation, long a concern of U.S. farming practice (McDonald 1941) gained national prominence. In 1935 the Soil Conservation Act was passed (Steiner 1990: xvi). At the end of the decade, the focus on conservation was again diverted by war.

The Postwar Conservation Movement
The postwar conservation movement focused on the "multiple use" of national forests. This involved a utilitarian calculus of trading off various uses, including recreation, grazing, logging, water supply, wildlife, fishing, and hunting. Under this approach, and absent any connection to the Progressive political tradition, conservation came to mean the technical management of natural resources. This discourse was increasingly replaced in the late 1950s and the early 1960s by the discourse of Environmentalism.

Sustainable Development
In the mid 1980s, conservation had a resurgence under the label of Sustainable Development. This term emerged with the publication in 1985 of the United Nations report *Our Common Future* (WCED 1987).[7] Sustainable Development is defined in this report as the management of natural resources in a manner that ensure that these resources "meet the needs of the present without compromising the ability for future generations to meet their own needs" (ibid.: 188). Since then, the real meaning of

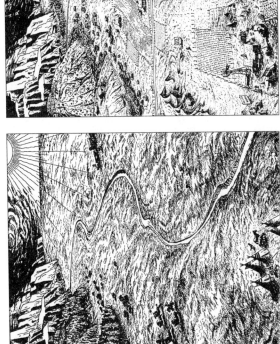

Figure 7.2
A unique illustration of the discourse of Conservation, from the report of the Great Plains Committee (U.S. Great Plains Committee 1936). The three sequential illustrations show the past, the present, and potential future of the Great Plains. At left, it is pristine, and full of wildlife. In the center, the results of unbridled exploitation and degradation are shown. This is seen to be the current state of these lands. At the right is the utopia of conservation, in which careful scientific management provides a bountiful and rich life.

sustainable development has been debated. There are many different definitions, none of which has achieved ascendancy. It represents a multivocal and ill-defined political goal that can be interpreted in many different ways (Goulet 1995: 44–59). In its essence, this idea is neither new nor novel. "The break-through of the much-touted Brundtland Commission," was, according to Evernden (1992: 76), "little more than Gifford Pinchot recycled for the nineties."

However, sustainable development has much of the same appeal that conservation had to the corporate and government communities. Torgerson (1995) maintains that Sustainable Development is essentially a discourse that attempts to ensure economic growth and to co-opt industrialism's environmental critics. It is "an incrementalist strategy that involves deliberate accommodation with established institutions orientated to the promotion of industrialism" (ibid.: 15). However, while it may be marginalist, and only a partial solution, it has led to the engagement of industry and the state in ecological issues (ibid.: 17; Torgerson 1999: 51–64).

One of the key strategies to emerge from the idea of sustainable development was *ecological modernization*, which seeks to reverse ecological degradation by using market mechanisms as instruments of an "ecologization of the economy" (Mol and Spaargaren 1993: 437). Through the use of new production technologies, ecological modernization is projected to bring about cleaner production processes, higher recycling rates, and lower emissions of pollutants (Torgerson 1999: 144).

Ecological modernization has been institutionalized in the industrial standards known as ISO 14000, which specify a series of practices a firm must follow to protect the environment. In some European countries, these standards have become mandatory. In addition, the academic discipline of industrial ecology is now emerging. This discipline studies how to integrate environmental concerns into the design, maintenance, and operation of industrial facilities. Thus, although sustainable development may be a latter-day version of conservation, it has interjected ecological concerns into the practices of industry.

Current Conservation Organizations
Organizations based on the discourse of Conservation define their objective as conserving or rationally developing our natural resources to meet long

term human needs. They use phrases such as "ensure wise use of natural resources" and "bring about efficient conservation and development." Their publications focus on successful instances of rational management of natural resources and on the problems that occur when this practice is not followed.

To examine the characteristics of organizations informed by this discourse, I compiled and examined a sample of eight Conservation groups. Based on this sample, the following general comments can be made.

The founding dates range from 1875 (American Forests) to 1985 (Rails to Trails Conservancy). The mean founding date for this sample is 1957. These organizations are of moderate size in terms of membership, staff, and income. The average group has 133,000 members and a full-time staff of 44.

The annual average income is $7,489,990, and the average of the groups' assets is $5,744,842. The sources of income of these groups are shown in table 7.2. Economic support comes primarily from members, with a significant component of foundation support. Accordingly, these groups have the potential to be influenced by external funding organizations.

As means of influencing policy, 57 percent of the groups follow primarily an education strategy, and 43 percent follow primarily a parliamentarian strategy. Thus, the strategies for social change are exclusively traditional and nonconfrontational. The organizational practices informed by the discourse of conservation are mixed, with some authentic grassroots organizations and some Astroturf organizations. Sixty-seven percent of these groups have a either a representative or a limited democratic structure.

Table 7.2
Sources of income of conservation groups.

	Percentage of income
Public contributions	24.4
Foundation grants	25.4
Government grants	2.5
Program services	18.4
Membership dues	22.3
Other	7.0

Thus, the organizational structure of these groups is mixed, with a slight trend toward more participatory structures. Those organizations without participatory structures have the potential to be less authentic community representatives. Thus, the discourse of Conservation provides some capacity for authentic movement organizations but does not strongly encourage their formation.

Social Learning and Conservation

Conservation is one of the most widely applied environmental discourses. Virtually every environmental problem has been viewed through this discourse. Thus, it is a general discourse that can encompass a wide variety of issues. However, it has several limitations.

First, it is a thoroughly anthropocentric discourse. It focuses on the natural world and its value for human activities. Other beings are not considered worthy of consideration in their own right. Thus, conservation cannot provide a basis for the protection of aspects of the natural world that do not serve human purposes. Hence, this discourse cannot inform a cultural practice that could protect biodiversity. As a result, conservation "translates questions about nature into questions of economic utility" (Taylor 1992: 57). "Thus," Taylor continues, "wilderness, with no significant market value, cannot be defended from this viewpoint."

Second, the discourse of Conservation has gradually lost its connection to the larger Progressive political tradition from which it originated. Without this connection to a larger notion of a just social order, conservation has been reduced to the technical management of natural resources in service of the existing social structure. This limits its ability to inform a comprehensive political and social agenda that could provide a basis for the creation of a sustainable society (Taylor 1992: 136).

Third, Conservation was the dominant environmental discourse in the United States for nearly 100 years, and during that period ecological degradation accelerated. The conservation movement was unsuccessful in preventing this from occurring. As an effective environmental discourse, Conservation has failed the test of time (Taylor 1992: xi). "Having tried the Pinchot experiment for the better part of a century, and having during that period sunk deeper and deeper into environmental despair," Evernden

(1992: 76) notes, "we still are not ready to conceded that perhaps the experiment has failed. How much longer must it be tested before we conclude that it is not enough?"

Preservation

In wildness is the preservation of the world.
—Henry David Thoreau (1862: 30)

If a war of races should occur between the wild beasts and Lord Man, I would be tempted to sympathize with the bears.
—John Muir (1912: 324)

The third environmental discourse to emerge in the United States during the nineteenth century was Preservation. Preservation defines a spiritual and psychological relationship between humans and the natural environment (McCormick 1989: 2; Oelschlaeger 1991: 289–292; Nash 1967: 57, 60). In this discourse, nature in the form of wilderness, untouched by human activity, has intrinsic value. Nature also serves as a site for self-renewal through the experience of its aesthetic beauty. This translates into a concern over the preservation of scenic areas, wilderness, and wildlife. The following are key components of this perspective:

• Natural systems are self-creating evolutionary wholes that cannot be reduced to the sum of their parts. Hence, nature is not a machine but an intact organism.
• Human actions can impair the ability of natural systems to maintain themselves or to evolve further.
• Wilderness and wildlife are important component in supporting both the physical and the spiritual life of humans.
• Human values go beyond those measured by the national income accounts to include the preservation of wild lands and life.
• Continued existence of wilderness and wildlife is critical to the spiritual well being of humanity.
• Protection of wilderness areas and wildlife for the current and future generations is an essential environmental task.

This discourse had a long period of gestation. The rise to dominance of a rational, Cartesian worldview in Europe had been long protested by the

Romantic movement, especially in Germany. It was also protested in the United States. An excellent example appeared in 1782 in *Letters from an American Farmer* by Hector St. John de Crevecoeur. "In contrast to this image of America as a fertile expanse to be developed by man," Worster writes (1973: 1), "Crevecoeur also perceived in American nature an unspoiled moral order. Humble conformity to this natural order would redeem a man's life from the corrupting artificiality of civilization. By following a simple, subsistence economy the American farmer could establish himself in ecological harmony with the land, recover an innocence that was lost to Europeans, and reawaken a sense of kinship with his fellow creatures." This viewpoint, combined with the Arcadian Myth of the inherent goodness of pastoral rural life over urban life, formed the historical position for the further development of the Preservation discourse in the United States (McCormick 1989 Schmitt 1969). Its origins can be traced through the transcendentalism of Emerson to the writings of Thoreau. By 1854, when Thoreau wrote *Walden*, the identity of this discourse was firmly established (Oelschlaeger 1991: 133–171). Thoreau sought to develop an alternative notion of the good life in opposition to a commercial and industrial society. In nature Thoreau found a vantage point for "criticizing both the superficiality and the downright evils of American society" (Taylor 1992: 15). He also found the inspiration for an alternative social order informed by an understanding of nature. Combining the image of a return to the Garden of Eden with the Arcadian Myth, Thoreau developed a vision of the good life as one lived with "simplicity and humility in order to restore man to a peaceful coexistence with nature" (McCormick 1989: 2). As summarized by the poet William Cullen Bryant, the message of nature suggested "the idea of unity and immensity, and abstracting the mind from the associations of human agency, carried it up to the idea of a mightier power, and to the great mystery of the origin of things" (Matthews 1991: 105).

This alternative vision, inspired and instructed by the natural world, had great political power and led to a significant social movement to create a better political community that would live in harmony with the natural world (Taylor 1992: 136). The development of a social movement based on this discourse originated in two areas. First, with the rise of urbanization and the associated problems documented by the muckrakers of the late nineteenth century, there arose a quest for wilderness as an alternative to urban civilization (Nash 1967: 57). "As a result of this sense of discontent with civilization," Nash (1967: 108) writes, "no less uncomfortable because of its vagueness, *fin de siecle* America was ripe for the appeal of the unciv- ilized on a broad popular basis." The second source of preservation was the rise in cultural esteem for natural beauty. In contrast to the cultural developments of Europe, the natural beauty of the United States was culti- vated as a source of unique American pride (Nash 1967: 67–83). The devel- opment of this attitude toward nature was assisted by the magnificent portrayals of the western landscape by painters such as Albert Bierstadt (National Gallery of Art 1992). To be able to enjoy this beauty was seen to be the mark of a refined person (Nash 1967: 60). The idea of self-renewal in the wilderness also led to a quest for a rural and genteel life. This quest animated the development of recreational uses of wilderness, such as hik- ing and camping. These activities translated into a concern for the preser- vation of wilderness areas and wildlife.

One of the earliest calls for the preservation of nature was made by the painter George Catlin in 1832:

And what a splendid contemplation . . . when one imagines them as they might in future be seen (by some great protecting policy of government) preserved in their pristine beauty and wildness, in a magnificent park, where the world could see for ages to come, the native Indian in his classic attire, galloping his wild horse, with sinewy bow, and shield and lance, amid the fleeting herds of elks and buffaloes. What a beautiful and thrilling specimen for America to preserve and hold up to the view of her refined citizens and the world, in future ages. (Catlin 1833: 235)

Thoreau (1858: 317) echoed this sentiment when he called for "our own national preserves . . . for inspiration and our own true re-creation."

This discourse was focused by John Muir. He kept the radical political critique of industrial civilization; however, instead of seeing nature as the source of an alternative vision of a social order, he saw wilderness as a source of religious inspiration and personal satisfaction (Taylor 1992: 104). Based on this perspective, the arguments for the preservation of nature by this discursive frame were primarily aesthetic. Natural areas were preserved for individual recreation involving spiritual renewal by exposure to natural beauty.

Beginnings of the Preservation Movement

The development of urban parks originated in the Scenic Cemetery movement of 1825–1850 (Huth 1948: 62; Huth 1957: 66–69). In 1825, a Boston physician, Jacob Bigelow, argued for the establishment of cemeteries outside cities for hygienic reasons. This cause was taken up by the Massachusetts Horticultural Society, and the first scenic cemetery—Mount Auburn, in Cambridge—was established in 1831. This cemetery was "was a spot of natural beauty which had been selected with the intention of conserving its original aspect" (Huth 1957: 67). It became a popular destination for weekend visitors. Scenic cemeteries were founded in New Haven in 1833 and in New York and Philadelphia in 1836.

The idea of an urban park in New York was first proposed by William Cullen Bryant in 1844. In 1849, the landscape architect Andrew Downing also advocated the development of urban parks. In July of 1849, citing the Scenic Cemeteries movement, Downing asked: "If 30,000 persons visit a cemetery in a single season, would not a large public garden be equally a matter of curious investigation?" (Huth 1957: 69) In 1851 this idea was realized when New York's City Council passed a resolution to create Central Park. Frederick Law Olmsted was appointed as the park's designer and its first superintendent (Huth 1948: 62; Clepper 1966: 150). In designing Central Park, Olmsted announced that he would create "a specimen of God's handiwork that shall be . . . inexpensively, what a month or two in the White Mountains, or the Adirondacks is" (Huth 1957: 70). Many parks were established in major U.S. cities in the second half of the nineteenth century (Huth 1957: 166).

In 1864, California designated Yo-Semite[8] Valley as a state park. The development of this park was precipitated by an incident known as the Calaveras Tree Murder (Huth 1948: 64). In the early 1850s, white settlers moving into California learned of the Giant Sequoia trees in the Sierra Nevada Mountains. Travelers came to marvel at these immense and ancient trees. One tree attracted special attention and was named the Mother of the Forest. This tree, the largest in the Calaveras Grove, was approximately 325 feet high and 30 feet in diameter. Estimated to be more than 2500 years old (Neuzil and Kovarik, 56–57), it was substantially larger than the largest tree in Kings Canyon National Park (the 274-foot-high General Sherman Tree) and the largest in Yosemite (the 290-foot Grizzly Giant). The stream of visitors to see these trees convinced an entrepreneur named George Gale that there was money to be made by exhibiting such a tree throughout the world, and in 1852, over the course of a month, Gale and his work crew cut down the Mother of the Forest. In *Harpers Weekly* (June 5, 1858: 357) the felling of the tree was described as follows:

This monster was cut down by boring with long and powerful augers, and sawing the spaces between them. . . . It required the labor of five men twenty-five days to effect its fall, the tree standing so near perpendicular that the aid of wedges and a battering ram was necessary to complete the desecration. But even then the immense mass resisted all efforts to overthrow it until in the dead of a tempestuous night it began to groan and sway in the storm like an expiring giant, and it succumbed at last to the elements, which alone could complete from above what the human ants had commenced below. Its fall was like the shock of an earthquake and was heard fifteen miles away. . . . Such was its vitality that, although completely girdled and deprived of its means of sustenance, it annually put forth green leaves until the past year.

The trunk was then cut up, shipped to the East Coast, reassembled, and exhibited in the United States and in Europe. The enterprise was an economic failure. Many of the viewers, especially in Europe, refused to believe that such a tree could exist, and dismissed the exhibition as humbug. Those who believed the exhibit to be authentic were so outraged at the destruction of such a wonder of nature for economic gain that they refused to go to the exhibit.

This incident led to calls for protection of trees and wilderness areas. In 1857 the editor of the *Atlantic Monthly* proposed a society "for the prevention of cruelty to trees" (Huth 1948: 63). This was followed by a number of articles and pictorial representations of Yo-Semite Valley (Huth 1948: 64–65). A proposal to protect Yo-Semite and the Mariposa Tree Grove was developed by a number of California residents, and in February of 1864 a California businessman named Israel Raymond forwarded the proposal to John Conness, one of California's U.S. senators. "I think," wrote Raymond, "it is important to obtain the proprietorship soon, to prevent occupation and especially to preserve the trees in the valley from destruction." (Huth 1948: 67) The driving force behind development of this proposal was Frederick Law Olmsted (Huth 1957: 149).

On May 17, 1864, Senator Conness introduced a bill to grant Yo-Semite Valley and the Mariposa Tree Grove to the state of California for the creation of a park (Newzil and Kovarik 1996: 53–82). In his speech, Conness cited the destruction of the Mother of the Forest and concluded: "The purpose of this bill is to preserve one of these groves from devastation and injury. The necessity of taking early possession and care of these great wonders can easily be seen and understood." (*Congressional Globe*, May 17, 1864: 2301) This bill, The Yo-Semite Valley and the Mariposa Big Tree Grove Land Grant, was signed into law on June 30, 1864, by President Abraham Lincoln. It granted Yo-Semite Valley to California "with the stipulation that the said state shall accept this grant upon the express conditions that the premises shall be held for public use, resort, and recreation" (Statutes of the United States, 38th Congress: 325). In September of 1864, the governor of California proclaimed the creation of Yo-Semite Park and appointed Olmsted as the first chairman of the Yo-Semite Park Commission. With the enactment of this law, the United States took its first step toward the preservation of wilderness areas. Yellowstone Park was established six years later as "a public park or pleasuring ground for the benefit and enjoyment of the people" (Statutes of the United States, March 1, 1872, chapter 24, section 1, 17).

Early Preservationist Groups
The social movement based on the Preservation discourse originated in mountaineering clubs. Perhaps the first American mountaineering club was the Alpine Club of Williamstown, Massachusetts, founded in 1863 (Scudder 1884: 45). This club's objective was "to explore the interesting places in the vicinity, to become acquainted . . . with the natural history of the localities, and also to improve the pedestrian powers of the members" (ibid.: 46). The Appalachian Mountain Club was founded in 1876, its purpose being to "explore the mountains of New England and the adjacent regions, both for scientific and artistic purposes; and, in general, to cultivate an interest in geographical studies" (quoted in *Appalachian Mountain Register* for 1948).

In 1892 John Muir founded the Sierra Club, whose purposes were "to explore, enjoy, and preserve the Sierra Nevada and other scenic resources of the United States, and its forests, waters, wildlife, and wilderness; to undertake and to publish scientific, literary, and educational studies, to educate the people with regard to the national and state forests, parks, monuments, and other natural resources of especial scenic beauty and to enlist public interest and cooperation in protecting them" (Sierra Club Articles of Incorporation, June 4, 1892). By the turn of the century, there were several similar organizations.[9]

There were also "fresh air charity" organizations, founded to provide city children with recreational outdoor activities. Concentrated on the East Coast, these groups took children on day and week trips to parks and to the shore. In an 1897 Ph.D. dissertation, one Walter Shepard Ufford wrote the following about these organizations: "The rapid development of our cities had sacrificed light and air to the pressing demands of modern industrialism. . . . Contemporaneously with the massing of population in the cities has come, like the memory of a departed blessing, a hunger for the sea, the country and the mountains. . . . Happily those who regularly take themselves off to the country . . . have not been able entirely to forget the less fortunate denizens of the city who are left behind." (Ufford 1897: 1) Ufford notes that 37 Fresh Air Societies were founded between 1870 and 1894. This tradition lives on in the Fresh Air Fund, founded in 1877.

Hetch Hetchy and the Split of Conservation and Preservation

Up until the end of the first decade of the twentieth century, the work of the conservationist and preservationist social movements had been joined, and John Muir had been friendly with Gifford Pinchot. However, a proposal to dam the Hetch Hetchy Valley for the purpose of creating a more secure source of water for San Francisco after the earthquake of 1906 led to a split between the two movements.

Pinchot and the conservationist groups saw the damming of this valley as an appropriate use of natural resources to meet human needs. In the same camp were the Cosmos Club (a group of San Francisco businessmen), a number of western senators, and William Randolph Hearst, editor and publisher of the *San Francisco Examiner* (City and County of San Francisco: 1994: 26–27). The opposition to the dam was led by Muir and the Sierra Club. To the preservationists, damming a beautiful and spectacular wilderness was sacrificing God's creation for human needs. Joined in the struggle to prevent the dam were mountaineering groups, wildlife conservationists, the American Civic Association, and the General Federation of Women's Clubs (Fox 1981: 143). Because the Sierra Club would sometimes waver in its opposition to the dam, Muir and his allies started an ad hoc group known as the Society for the Preservation of National Parks (ibid.: 144).

In 1913 the struggle over Hetch Hetchy culminated when Congress voted to construct the dam. The Sierra Club and its allies sought a more secure legislative basis for the National Parks, and in 1916 a new law defining the activities that may take place in a National Park and establishing a National Park Service was signed into law by President Wilson. The conservationists and the preservationists remain divided to this day. The Sierra Club is still intent on removing the Hetch Hetchy dam, and a task force is working to develop and advocate proposals.

The Pursuit and the Defense of Wilderness

Just before World War I, and after the war, the National Forests were used increasingly as recreational areas (Baldwin 1972: 11–12). Recreational use began to compete with mining, logging, grazing, and development of the National Forests. To balance these uses, the National Forest Service began to develop comprehensive management plans.

In developing the plan for the White River National Forest in Colorado in 1919, a forest service employee named Arthur Carhart suggested that a certain portion of the land around Trappers Lake should remain roadless and undeveloped. This designation was approved, and Trappers Lake became the first de facto wild area in the National Forests (Baldwin 1972: 29–30). Carhart also sent a memo to Assistant Forester Aldo Leopold advocating the development of a Forest Service policy that would specify certain areas "where all such developments . . . shall stop" (Carhart 1919). "These areas," he argued, "can never be restored to the original condition after man has invaded them. . . . Time will come when these scenic spots where nature has been allowed to remain unmarred, will be some of the most highly prized scenic features of the country." Carhart's idea took hold, and in 1922 a portion of the Superior National Forest in Minnesota was designated as a wilderness area. A third wilderness area, in New Mexico's Gila National Forest, was added under plans developed by Leopold in 1924. In 1929, formal criteria for the establishment of "Primitive Areas" were issued by the Chief of the National Forest Service (Carhart 1972: 200–204).

Some did not feel that the existing Forest Service regulations were adequate to protect wilderness areas throughout the United States. Writing in

1930, Bob Marshall, then a U.S. Forest Service employee, argued for collective action to preserve wilderness: "There is just one hope of repulsing the tyrannical ambition of civilization to conquer every niche on the whole Earth. That hope is the organization of spirited people who will fight for the freedom of the wilderness." (Marshall 1930: 11) Marshall became an outspoken advocate of wilderness. In his 1933 book *The People's Forests*, he argued for the socialization of all forested property. According to Fox (1981: 207), Marshall argued "that the history of utility regulation in the United States had largely resulted in private interests regulating regulation, and that private ownership in its focus on immediate profits could not provide the continuity essential to forest management. He concluded that every acre of woodland in the country should belong to the people."

Marshall's 1930 call for action came to fruition in 1935 when eight individuals, including Aldo Leopold and Bob Marshall, founded the Wilderness Society. The opening article in the society's newsletter *The Living Wilderness* (volume 1, no. 1, September 1935)—titled A Summons to Save the Wilderness—asserts that "wilderness is a natural mental resource having the same basic relation to man's ultimate thought and culture as coal, timber, and other physical resources have to his material needs. In the same issue, the purpose of the new society is specified clearly: "The Wilderness Society is born of an emergency in conservation which admits of no delay. It consists of persons distressed by the exceedingly swift passage of wilderness . . . and who purpose is to do all they can to safeguard what is left of it."

Little preservation activity occurred during World War II, and by that war's end the preservationist movement had become comfortable with the status quo. Most of the nation's policies regarding National Parks and wildlife were formed with the cooperation of wildlife-management organizations, such as the Audubon Society and the Izaak Walton League. The National Wildlife Federation, which focused on education and on membership services, did not live up to its founders' desire to be a strong voice for conservation (Fox 1981: 262). Thus, the movement consisted of comfortable insiders, politically ineffectual and isolated from their grassroots support. As described by Fox (ibid.: 250), it was "an establishment based in Washington and New York . . . controlled most of the available money and personnel. Within this establishment the key individuals knew each other well and met . . . at the Cosmos Club in Washington. . . . This inbreeding of leadership facilitated cooperation but, inevitably, exacted a price in isolation from the hinterlands."

In this void of leadership and action, the Sierra Club transformed itself from a regional group into one of the nation's leading preservationist organizations. Since the Hetch Hetchy battle, the club had emphasized outings and hiking at the expense of political action. In the late 1940s, under the leadership of David Brower, a group of dedicated and militant insurgents took control of the club. Wilderness advocacy expanded, and the *Sierra Club Bulletin* shifted from being a hiking guide to a more political publication with a focus on preservation battles (Fox 1981: 278–279). The club's membership and its budget expanded rapidly.

The Sierra Club emerged as a leading national preservation group in the period 1950–1956 with its fight to stop a dam from being built in Dinosaur National Monument. Under Brower's leadership, a coalition of 17 groups, including the Audubon Society, the Wilderness Society, and the Izaak Walton League, was formed. Drawing parallels to the damming of Hetch Hetchy 40 years earlier, this coalition successfully challenged the Eisenhower administration's plans for the Echo Park dam (Fox 1981: 272–281).

In 1957, building on the coalition's success in the Echo Park dam fight, Howard Zahniser of the Wilderness Society began a campaign that resulted in the Wilderness Act (P.L. 88-577). Signed into law in 1964, this act provided strong legislative protection for wilderness. Its purpose being to "secure for the American people of present and future generations the benefits of an enduring resource of wilderness," the Wilderness Act defined wilderness as "an area where the Earth and its community of life are untrammeled by man, where man himself is a visitor who does not remain."

The Wilderness Act was followed in 1968 by the Wild and Scenic Rivers Act. Later victories for preservation include the Endangered Species Act (1973) and the Alaska National Interest Lands Conservation Act (1980). The main focus of preservationist groups today is reauthorization of the Endangered Species Act.

Current Preservationist Organizations
Organizations based on the discourse of Preservation define their objective as preserving wilderness in a pristine state, untouched by humans. This

includes leaving all the plants and wildlife that inhabit that area to develop in a "natural" manner, unaffected by human influences. They use phrases such as "preserve and protect" and "ensure the continued existence of wilderness areas."

Some preservationist organizations focus only on a specific species or geographic region. This is reflected in names such as Save the Whales, Save the Sea Otter, and Mono Lake Committee.

The magazines produced by preservationist organizations celebrate wilderness and wildlife through the use of extraordinary pictures and descriptive articles.[10] The magazines describe threats to wilderness areas or wildlife and the need to take action to preserve them, and they report to their membership on the triumphs that the organization has had in preserving a particular species or wilderness area.

To examine the characteristics of organizations informed by this discourse, I compiled and examined a sample of eighteen preservationist groups, all founded either around 1900 or between 1960 and 1975. The mean founding date is 1933. Most of these organizations are moderate in size, though the Sierra Club, the Wilderness Society, and the Nature Conservancy are extremely large. The average membership is 248,000, the average staff 235.

These groups have an average annual income of more than $34 million and hold average assets of $75 million. One organization, the Nature Conservancy, had $321 million in income and $1.03 billion in assets.[11] No other group in all of the U.S. environmental movement even approaches these economic resources. The sources of income are listed in table 7.3. Some preservationist groups derive a large portion of their income from

Table 7.3
Sources of income of preservation groups.

	Percentage of income
Public contributions	40.8
Foundation grants	15.3
Government grants	6.7
Program services	8.4
Membership dues	8.8
Other	20.0

corporate support. (Unfortunately, corporate sponsorship is not differentiated in the data.) The Nature Conservancy, which has dozens of corporate sponsors, derives nearly 40 percent of its income from foundation grants.

In strategies for influencing policy, these groups are very similar to Conservation groups. Fifty-three percent follow primarily an education strategy, 47 percent primarily a parliamentarian strategy. Sixty-three percent of preservationist groups are oligarchic, 25 percent have limited democracy, and 12 percent are democratic.

Social Learning and Preservationism

Preservation is a limited discourse that focuses on the protection of wilderness and wildlife. This focus limits the range of issues and concerns addressed by preservationist organizations. As a result, these organizations have been accused of attending only to the recreational needs and aesthetic values of middle- and upper-class whites.

These organizations engage in traditional political activities and do not pursue wide-ranging social change to protect the natural environment. In addition, they have abandoned Thoreau's radical critique of industrial civilization and his alternative vision of the good life for a limited effort to save wilderness for its own sake. This limited vision hinders the development of an alternative cultural model that can inform an ecologically sustainable society. As Taylor (1992: 136) notes, "contemporary pastoralists have largely been unable to derive a coherent alternative politics and social vision from their deep respect for nature."

Owing to their high incidence of oligarchic structure, preservationist organizations have a history of becoming isolated from the members they supposedly represent. In addition, the significant economic support they receive from foundations indicates that these groups have a substantial potential to be channeled and co-opted by external funding organizations. Together, these two characteristics suggest that they have a high potential to become "Astroturf" organizations.

8

Reform Environmentalism: Public Health and Ecology

The various forms of epidemic, endemic, and other disease caused . . . amongst the laboring classes (are from) atmospheric impurities produced by decomposing animal and vegetable substances, by damp and filth, and close and overcrowded dwellings.
—Edwin Chadwick (1842: 369)

We have broken out of the circle of life, converting its endless cycles into man-made, linear events. . . . Suddenly we have discovered what we should have known long before: that the ecosphere sustains people and everything that they do.
—Barry Commoner (1971: 12)

The most dominant environmental discourse of the present day is Reform Environmentalism. This discourse is, in fact, so dominant that it is generally used to refer to the multiple discourses and communities that make up the current environmental movement.

Founded on the natural sciences and the notion of humanity as part of the Earth's ecosystems, the discourse of Reform Environmentalism links human survival to environmental conditions (Oelschlaeger 1991: 292–301). In this discourse, nature has a delicate balance, and humans are part of it. This perspective emphasizes that nature is an ecological system, that is, a web of interdependent relationships. Humanity is part of this ecological system. Hence, human health is vulnerable to disturbances in the ecosystem. This animates action to identify and eliminate the physical causes of environmental degradation.

The following are key components of this perspective:

• Natural systems are the basis of all organic existence, including that of humans.
• Humankind is an element within natural ecosystems, and hence human survival is linked to ecosystem survival.

• Ethical human actions (actions that promote the good life for humankind) necessarily promote action toward all life on Earth in an ecologically responsible manner.
• Proper use of the natural sciences can guide the relationship between humanity and its natural environment.

This discourse originated in the same historical era as both Conservation and Preservation. However, not until the mid 1960s did it became a leading environmental discourse.

The Origins of Reform Environmentalism

Reform environmentalism is a complex system of beliefs whose origins lie in the utilitarian philosophy of providing for the common good through the application of science and law to public problems and in Thomas Malthus's writings on overpopulation. These two perspectives inform different components of the current environmental movement.

Major elements of environmentalist beliefs can be traced to the early nineteenth century, specifically to the works of Jeremy Bentham and the idea of utilitarianism. In utilitarianism, a good action was one that benefited the material interests of the community. Bentham (1973: 18) defined a concept known as the *principle of utility*: "An action . . . may be said to be conformable to the principle of utility . . . when the tendency it has to augment the happiness of the community is greater than any it has to diminish it." Based on this perspective, Bentham and his followers, the Philosophic Radicals, developed a program of social reform. They proposed the use of rational science to address questions of public policy. Through rational science, the most utilitarian policy could be developed. This policy and the application of the principle of utility should then be the primary guide to action of state legislators.

One of the areas in which this philosophy was applied was in developing ways to address the emerging urban environmental problems. Summarizing the work of the utilitarian philosophers, Petulla (1987: 17) maintains that they were concerned with "environmental causes of poverty and disease, and how scientific/utilitarian policies could promote more efficient government and industry," and that they "believed that these reforms could be implemented most effectively on a national level by Parliament."

This theory was applied to the problem of public health in the 1840s in Britain. At that time, the miasma theory of disease was current. This theory connected the causes of diseases to the state of the atmosphere. Foul odors were seen to be disease-carrying miasma. The key to preventing disease was to remove the sources of the foul odors, such as excrement, decaying organic matter, or filthy streets (Petulla 1987: 16–17). This scientific theory was applied to the problem of urban health by the disciples of Bentham, the most influential of whom was Edwin Chadwick. Chadwick maintained that exposure to the decay and filth of a slum created high levels of disease among the poor. Thus, it was poverty and poor sanitation that were the causes of disease. This linked both the social and the physical environment to human health.

Chadwick was commissioned to report on the causes of disease. His Report on an Inquiry into the Sanitary Conditions of the Laboring Population of Great Britain was published in 1842. The most important characteristic of Chadwick's report was that it linked environmental conditions and human health (Melosi 1981: 12). Specifically, it connected sickness and disease to filth, pollution, and polluted water supplies among the people who lived in conditions of poverty. It recommended a program of sewers and development of sanitary water supplies (Petulla 1987: 17–20). This report formed the basis for the development of sewers in Britain. It also established the model for environmental inquiry. It contained a wealth of data, analyses of economic and environmental conditions, and recommendations for further legislative and technical actions to alleviate the problem.

In his famous *Essay on the Principle of Population*, published in 1798, Malthus argued that the primary threat to human survival was overpopulation. He based his argument on the idea that food supplies developed linearly while the human population increased geometrically. Thus, shortage and human misery would always be with us. To alleviate misery for all, Malthus argued, there must be a maximum effort in the production of food, there must be moral restraint in sexual activities to reduce population growth, and the Poor Laws (which provided sustenance to the poor) should be abolished (Paehlke 1989: 43). Malthus's ideas had no large political impact until several writers revived them in the late 1960s. These writings subsequently inspired the development of several groups dedicated to the control of human population growth.

However, the writings of Chadwick and the early utilitarians gave rise to a host of efforts to improve urban environmental conditions almost immediately after their publication.

The Beginning of Environmental Concerns

Industrialization brought the burning of coal, the concentration of factories, and human crowding in urban areas. This created environmental problems in the industrial cities in the United States. "Crowded tenement districts, chronic health problems, billowing smoke, polluted waterways, traffic congestion, unbearable noise and mounds of putrefying garbage" resulted in severe urban health problems and in high rates of disease and mortality (Melosi 1981: 16–19). The brunt of the extraordinary deprivations of nineteenth-century urban life was borne by the working class and the poor (ibid.: 10). These conditions created concerns about urban environmental conditions in three areas: sewers and water supplies, sanitary streets, and clean air.

Sewers and Clean Water: The Sanitary Movement

Chadwick's report on the sanitary conditions in England was almost immediately followed by a similar study in the United States: The Sanitary Condition of the Laboring Population of New York (Griscom 1845). This report, which also linked disease to the squalor and filth of tenement living of the poor (Petulla 1987: 23), fostered public concern about the deteriorating air and water quality of urban areas. These environmental conditions were seen to be the cause of epidemics of typhoid, dysentery, cholera, and yellow fever (ibid.: 24).

After the Civil War, there was a rising concern with urban living conditions and their effects on public health among the middle class residing in urban areas. It was generally accepted at that time that unsanitary conditions caused disease (Rosen 1958: 240–286; Tarr 1985: 520). Hence, the foremost concern of urban reformers was the development of an infrastructure to protect human health (Melosi 1985: 494–515). This concern over the urban environment translated into political demands for action to address these problems, which took the form of the Sanitary Movement (Tarr 1985: 520).

Among the concerns of the Sanitary Movement were sanitary water supplies, sewage systems, garbage, and air pollution (Tarr 1985: 516–552).

One of the early groups associated with this movement was the New York Sanitary Association. Founded in 1859, its aim was "the improvement of the sanitary condition of the people, and, so far as connected therewith, the advancement of their economic and moral interests . . . by promoting the investigation of facts and principles relating to personal, domiciliary, and public hygiene" and "by diffusing information on the laws of health and life, and the best means for their application" (New York Sanitary Association 1859: 9–10).

Other such groups founded throughout the United States sponsored studies of hygienic conditions and lobbied for improvements. A report prepared by the Citizens' Association of New York in 1866 examined a wide range of conditions, including tenant housing, drainage and sewerage of the city, disease rates, and special nuisances to city health. The nuisances included "noisome gases from manufactories of various kinds" (Council of Hygiene and Public Health, Citizens' Association of New York 1866).

The Sanitary Movement created significant political pressure to build water and sewer systems. In 1860, 136 cities had some form of water system. By 1880, there were 598 sewer and water systems in operation. The last major U.S. city to get sewage and water systems was Baltimore, in 1911 (Petulla 1987: 24; Tarr 1984: 7). The dominant model for water systems was filtering and cleaning of incoming water, and discharge of sewage into rivers, allowing downstream flow to dilute and disperse the sewage (Petulla 1987: 28). Disputes arose between physicians and sanitary engineers on whether there was a need to treat the sewage before it was discharged into streams and rivers, or if dilution was adequate. The sanitary engineers prevailed, and raw sewage was generally dumped untreated into rivers and waterways. To ensure that the population was provided with clean water despite the continued discharge of sewage and industrial waste into the rivers, water filtration and chlorine disinfecting technologies were developed (ibid.: 29–30). This model of treatment remained essentially unchanged until the passage of the Clean Water Act of 1972, which limited discharge of sewage and industrial waste into waterways (Tarr 1984: 8–9).

Clean Streets and the Municipal Housekeeping Movement

Urbanization also created concern regarding the excess of refuse and garbage in urban areas. Several protests occurred, and a number of civic organizations were founded in the 1890s to demand urban cleanliness. This movement, which became known as the Municipal Housekeeping Movement, was composed primarily of educated middle-class women. In 1915, Mildred Chadsey, Cleveland's Commissioner of Housing and Sanitation, defined Municipal Housekeeping: "Housekeeping is the art of making the home clean, healthy, comfortable and attractive. Municipal housekeeping is the science of making the city clean, healthy, comfortable and attractive." (Chadsey 1915: 53) This movement primarily took the form of anti-littering campaigns, education about sanitary procedures, city cleanup days, and advocating effective sanitation ordinances (ibid.: 121–132).

A number of local and city organizations were founded to press for action by city officials. One of the most influential organizations of this movement was the Ladies' Health Protective Association of New York City, founded in 1884 (Melosi 1981: 34–35). This movement was a part of the Progressive Movement to make cities desirable places to live. A number of urban reform groups formed throughout the United States (ibid.: 106–107). In 1901 there were 11 states with urban reform organizations. By 1915, 29 states had urban reform groups. The urban reform movement emerged on the national level with the founding of the National Municipal League in 1894, the League of American Municipalities in 1897, and the National League of Improvement Associations in 1900 (ibid.: 108–112). By 1910 there were hundreds of such groups that could mobilize hundreds of thousands of members. One of their major efforts was to monitor urban areas for cleanliness (Petulla 1987: 24–25).

In 1894, Colonel George Waring, a veteran of the Civil War, was appointed New York City's Street Cleaning Commissioner. Applying his military experience to the task, he developed a military-like street cleaning corps. The street cleaners came to be called the White Wings, after the distinctive hats that they wore with their white uniforms (Melosi 1977: 59–78).

Colonel Waring, a tireless crusader for urban sanitation efforts, worked closely with a number of citizens' groups such as the Good Government Club, the City Improvement Society, and the Ladies' Health Protective Association (Melosi 1981: 73). He also organized school children into Juvenile Street Cleaning Leagues. By 1899, there were 75 of these leagues in the city, with more than 5000 participants (ibid.: 74–75).

Waring and his army of street cleaners succeeded in clearing the streets of New York of trash and snow. Waring also developed several innovations in handling solid waste, including the first rubbish-sorting plan in the United States. This enabled the development of recycling, including the use of ashes and organic materials to build bricks and the extraction of grease and other materials for use as fertilizer. He also developed the first sanitary landfills (ibid.: 72–73; Petulla 1987: 26).

To raise support and enthusiasm for urban sanitation efforts, Waring held parades of the White Wings through the streets of New York. The first such parade was held in May of 1896. More than 2000 sweepers and 23 bands paraded down Fifth Avenue, with Waring leading the parade, mounted on his prancing steed and meticulously dressed in his uniform, with pith helmet and riding boots (ibid.: 65–66).

Waring and his advocates were "able to produce a workable model for sanitary reform first and foremost because his message was clear: waste is a menace to health and to palatable living conditions and can be eradicated only through the coordination of municipal authorities and civic action" (ibid.: 77).

Early in the twentieth century the miasma theory was superseded by the germ theory of disease, and public health officials shifted their concern from sanitation to inoculation and immunization. Sanitation efforts were turned over to the technical and administrative control of sanitary engineers, and the days of the White Wings came to an end (ibid.: 80–104).

Although the sanitary movement was generally successful in dealing with urban refuse, it left several problems for future generations to solve. For one thing, landfills were seen as the solution to the problem of solid waste. Unfortunately, the landfills were not properly designed, and they left a legacy of pollution for future generations. In addition, the handling of toxic and hazardous wastes was generally not addressed before 1976. The legacy of these actions is evident in the numerous toxic-waste dumps throughout the United States (Tarr 1984: 16–26).

Clean Air and the Smoke Prevention Movement

By 1900 the burning of coal had severely affected the quality of urban air. Significant human health problems due to the breathing of smoke had been identified in several urban areas. Loggers had even noted the death of trees around large cities. These problems led to the creation of an alliance of civic and women's groups that came to be known as the Smoke Prevention Movement (Melosi 1981: 30–31).

One of the first organizations founded to promote the improvement of the urban environment was the Smoke Prevention Association of America (now the Air Pollution Control Association), founded in 1907. Its purpose was to "foster the control of atmospheric pollution and improve sanitation of the air by promoting the . . . abatement and/or prevention of atmospheric pollution affecting health and/or damage to property, nuisance to the public, and wasting natural resources" (by-laws, Smoke Prevention Association of America, 1951).

There were a number of efforts in large cities to reduce smoke from industrial activities. By 1912, 23 of the 28 U.S. cities with populations over 200,000 had some smoke-control ordinances (Tarr 1984: 10; Melosi 1980: 93). These local ordinances were mostly ineffectual. Further efforts were made during and after World War II to control urban smoke and air pollution, without much success (Tarr 1984: 14).

In the summer of 1907, a group of working-class women in Philadelphia appealed to the local civic club about urban noise:

What we can not stand is the noise. It never stops. It is killing us. . . . No one can sleep till midnight and all the noise begins again at five. Many of us have husbands who work all night and must get their sleep during the day, but they can get no sound sleep with all the noise that goes on about us. . . . Now what can your civic club do for us? (Melosi 1980: 139).

The civic club responded by forming a Committee on Unnecessary Noise to work on legislation to curb noise in the city. In 1912, a Society for the Suppression of Unnecessary Noise was formed in New York to limit the noise from boat whistles in the harbor (Melosi 1980: 141).

There were also organizations to promote personal sanitary hygiene. In 1911 the American Association for Promoting Hygiene and Public Baths was formed in New York to promote proper sanitation habits in children and the development of public baths (Proceedings of the American Association for Promoting Hygiene and Public Baths: 1916).

The Legacy of the Early Urban Environmental Movements
When the germ theory of disease replaced the miasma theory, public health was redefined as a medical specialty rather than a public issue. Hence, concern over the urban environment was redefined as a medical issue and a professional activity (O'Brien 1983: 13–15). Thus, the discourse of Reform

Environmentalism failed to develop into a widespread movement concerned with the quality of the urban environment, and it failed to connect with either the conservationist or the preservationist movement until the 1960s.

As the Progressive era ended with World War I, so too did the campaigns for the urban environment. There was some sporadic activity in the 1920s and the 1930s. The Izaak Walton League continued to press for clean water, since it affected fishing. There were also initial attempts to deal with the problem of oil spills. In 1922 the National Coast Anti-Pollution League was formed to "foster and aid the enactment and enforcement of adequate remedies and legislation to prevent the pollution of navigable and inland waters, and to secure the co-operation of those responsible for such pollution in accomplishing its elimination by all lawful means" (Phinney 1924: 9).

Academic research continued to examine the relationship between natural conditions and human health. An example of this type of research is contained in the proceedings of a symposium held at Harvard University in August of 1937. Titled "The Environment and Its Effect upon Man," this symposium covered a number of industrial and occupational safety topics, as well as air pollution concerns. However, there was no significant action regarding urban environmental problems until after World War II.

No unified reform environmental movement emerged from these distinct reform efforts. Assessing the overall effect of the early urban environmental movement, Melosi (1981: 132–133) notes that "each group possessed powerful reform tools, and each had links to important constituencies, but the gap between them was never bridged." Melosi continues: "Lacking a sufficiently broad environmental perspective, antismoke groups, noise abatement groups and sanitary reform groups pressed their specific causes independently, failing to coordinate their efforts." A more unified perspective would take another 50 years to emerge.

The Age of Reform Environmentalism

After World War II, the urban problems of pollution continued to expand. The early urban environmental movements had developed local solutions to their environmental problems. Sewage was sent downstream in the rivers, refuge and wastes were buried underground, and air pollution was lifted out of the local areas by tall smokestacks and allowed to dissipate into the larger atmosphere. By solving local problems, the early environmental solutions led to the creation of wider regional, national and global problems. The rivers became polluted and unsafe for swimming and drinking, landfills began to leach toxic chemicals, and acid rain developed as a result of the burning of coal and development of tall smokestacks.

In addition to the accentuation of these earlier problems, there appeared new problems related to post-World War II economic development. The dramatic growth of the petrochemical industry created a whole new class of pollutants, with widespread effects. The reliance on the automobile created a new form of urban air pollution known as photochemical smog. Finally, the development of mass consumption in the affluent society rapidly increased the use of resources and the production of waste.

As this situation developed in the postwar period, a number of books concerned with environmental conditions were published. Two that appeared in 1948—*Our Plundered Planet* by Fairfield Osborn and the best-seller *The Road to Survival* by William Vogt—revived Malthusian ideas. The concern about population raised by Vogt's book led to the creation of a government commission in 1952 to investigate the problems of overpopulation (Fox 1981: 306–312). In 1951, Rachel Carson published *The Sea around Us*, which increased concern regarding the ocean environment.

The administration of Dwight D. Eisenhower did little to deal with environmental pollution. However, with the inauguration of John F. Kennedy, there was an increase in executive leadership to deal with pollution. In his "Special Message to the Congress on Natural Resources" of February 23, 1961, Kennedy advocated the traditional conservation approach in the development of water projects, and appropriate management of federal lands. He also went beyond the traditional conservation message in advocating significant additions to the national parks. To carry out these efforts, he appointed Stewart Udall as Secretary of Interior. Udall worked to develop a series of major environmental initiatives (Caulfield 1989: 29).

Popular concern over the natural environment dramatically increased with the publication of Rachel Carson's *Silent Spring* in 1962. Historians are virtually unanimous about the importance of this book in development of modern reform environmentalism (Marcus 1985: 51; McCormick 1989:

47, 55–56; Dunlap and Mertig 1991: 2). This book reunited conservationist and preservationist concerns about natural ecosystems with long-dormant Public Health concerns (O'Brien 1983).

Silent Spring redefined the form of human-nature interaction and founded the modern Reform Environmental discourse. According to O'Brien (ibid.: 17–18):

This book crystallized two themes that were to be crucial to Reform Environmentalism: (1) nature has a delicate balance and (2) humans are part of it. . . . The first theme expressed the extent to which, for a large number of people, the appreciation of nature had broadened into a sense of ecology, that is, the web of interdependencies in particular natural environments. Carson's second theme—the vulnerability of humans to disturbances in the natural order—brought the ecological argument home. Carson argued that human health was being endangered by the overuse of new chemical pesticides, many of them resistant to the normal breaking-down processes of nature. The book was not only an exposition of ecology, but also a tract on public health; by weaving the two themes together, Carson gave them a force they would never have had by themselves.

In *Silent Spring* Carson also criticized the politics of science and the exclusion of the public from knowing what risks they were being exposed to by the development and use of synthetic chemicals. Foreshadowing the later work of Barry Commoner and Ulrich Beck, Carson's work exposed the public to the ongoing debate within the scientific community. Killingsworth and Palmer (1992: 65) comment that "the deep significance of *Silent Spring* lies less in its claims about the wrongdoings of the chemical industry, and more in its expose of ideological diversity within the scientific community." By involving the public in this debate, Carson focused public scrutiny on the need for "socializing research and control in the form of government regulation in matters where the public environment is concerned" (ibid.: 77). This interest was displayed in a speech that Carson gave before the Garden Club of America on January 8, 1963. In this speech, she expressed concern over the association between science and industry:

We see scientific societies acknowledging as "sustaining associates" a dozen or more giants of a related industry. When the scientific organization speaks, whose voice do we hear—that of science or the sustaining industry? . . . As you listen to the present controversy about pesticides, I recommend that you ask yourself "Who speaks?—and Why?" (Carson 1963: 15)

Concern over environmental deterioration due to industrial activities increased with the publication of Murray Bookchin's *Our Synthetic*

Environment, also in 1962. In 1963 Stewart Udall published *The Quiet Crisis*, which advocated a program for the protection and preservation of the natural environment (Caulfield 1989: 27–31). On the political front, the momentum for environmental issues also increased. Based on a coalition that had been built through the late 1950s, the Sierra Club reemerged as a strong national organization leading the environmental effort. David Brower took a major role in ensuring the engagement of the Sierra Club in new environmental issues of air and water pollution. With the Sierra Club working in conjunction with the other environmental-movement groups (especially the Wilderness Society) and with Secretary of the Interior Stewart Udall, the Wilderness Act was passed in 1964. The political momentum continued in the administration of Lyndon Johnson. A number of important environmental laws were passed, including the Water Quality Act in 1965, the Clean Air Act in 1967, the Wild and Scenic Rivers Act in 1968.

Out of these events the reform environmentalism of the late 1960s and the early 1970s was born. It was part of a broad-based social movement that sought to develop a cultural alternative to the modern social order (Hays 1985: 202; Hughs 1988). It also grew out of a personal response to the degradation of the individual lifeworld. Hays (1981: 721) argues that the idea that the origin of environmental activism is based on "intensely personal reactions to environmental quality and environmental degradation is unmistakable," and that "the emotion, the drive, the persistence of what is often called environmental activism stems from the meaning of environmental values at this primary group context of life."

Reform environmentalism had several direct links with the antiwar movement, the student movement, and the nuclear disarmament movement. In fact, many of the same people worked simultaneously for several of these movements (McCormick 1989: 61–64; Scheffer 1991). Copying the civil disobedience tactics developed by the civil rights movement, some environmental organizations, most notably Environmental Action (Mitchell et al. 1992: 20), joined in direct actions with other progressive social movements to challenge the social order (McCormick 1989: 62). This early appearance of radical environmentalism presaged the next such type of actions in the early 1980s. In addition, the appearance of Ralph Nader's book *Unsafe at Any Speed* in 1965 contributed to a demand for public accountability of government agencies that became part of the environmental agenda (Scheffer 1991: 25–26). David Zwick, one of Ralph Nader's staff workers, wrote a book on water pollution and then went on to found Clean Water Action.

The founding of the Environmental Defense Fund, in 1967, marked the beginning of a wave of new organizations based on the discourse of Reform Environmentalism. The EDF describes itself as "a nationwide public interest organization of lawyers, scientists, and economists dedicated to protecting and improving environmental quality and public health" (Gale 1990). It was founded following the successful actions of an informal team of lawyers and biologists working on behalf of the Audubon Society. This group brought a legal suit based on scientific research to halt spraying of DDT on Long Island. The resulting court action developed a new strategy in which scientific knowledge and legal action were linked to protect the environment. From this initial success, the Audubon Society, the Rockefeller Fund, and the Rachel Carson Fund provided the initial seed money to start the Environmental Defense Fund (Rogers 1990: 26–33; Ingram and Mann 1989: 137). The objective of EDF, according to its by-laws, is "to encourage and support the wise use of natural resources, and the maintenance and enhancement of environmental quality."

The pattern established by the Environmental Defense Fund consisted of using scientific research and legal action to protect the environment and human health. This organization serves as the exemplar for many other organizations that followed, including the Natural Resources Defense Council (whose slogan is "Law and Science in Defense of the Environment") and Friends of the Earth. The purpose of the latter organization— almost identical to that of the EDF—is to "combat and eliminate water pollution and air pollution and the ill health resulting from water pollution and air pollution, and to promote the wise management of natural resources for the betterment of mankind" (by-laws of Friends of the Earth, 1990).

One of the key intellectual leaders of reform environmentalism was Barry Commoner. In two highly influential books, *Science and Survival* (1966) and *The Closing Circle* (1971), he argued that the primary cause of ecological degradation was that economic forces had caused less-destructive productive technologies to be replaced by technologies with more intense environmental effects (Dunlap 1992: 79). Anticipating the arguments of Ulrich Beck,[1] Commoner focused on the social forces affecting the development of scientific knowledge, and the need for independent scientific knowledge and public input into the development of production technologies.

In 1968 Garret Hardin published an essay titled "The Tragedy of the Commons," the core idea of which was that environmental problems could be traced to the pressures caused by human overpopulation. In the same year, Paul Ehrlich, another key intellectual influence on the environmental movement, published *The Population Bomb*. Arguing that Malthus was essentially correct, Ehrlich argued for the development and enforcement of policies to restrict human population growth. To carry out these policies, a strong administrative state would be needed to institute and enforce population-growth restrictions (Paehlke 1989: 41–75). This theme has played itself out in modern environmental politics with the development of several groups dedicated to the control of population growth.

This phase of the environmental movement peaked on the first Earth Day, April 22, 1970, in the largest environmental demonstration in U.S. history, participated in by an estimated 20 million people (McCormick 1989: 47). The poster for this event featured a globe overflowing with people, many of them falling off. This symbolized the limits to growth of a finite Earth. There were many demonstrations held throughout the United States. University of Washington students put out a bucket of oil and invited onlookers to dip their hands in it so they would know how it felt to be a bird caught in an offshore oil slick. Students held "Trash Ins" or "Dump Ins" in which refuse was collected for return to the companies which originally produced it. Activists piled thousands of no-return bottles on the steps of Coca-Cola's headquarters. Students at Florida Tech held a trial to condemn a Chevrolet for poisoning the air and then symbolically buried the car. Students from Kent State planned a mock funeral for "the children of tomorrow," with a horse-drawn hearse to lead a procession of mourners through the streets. At the University of New Mexico in Albuquerque, students collected signatures on a big plastic globe to present as an "Enemy of the Earth" award to 28 state senators accused of weakening a recent anti-pollution law. San Francisco group dumped oil into a reflecting pool at the offices of the Standard Oil Company of California in a protest against oil slicks. (source: A Reflection on the Roots of Earth Day 1995, Earth Day Network, 1995)

Figure 8.1
A poster for the first Earth Day (from *Environmental Action*, April 1970).

The political momentum of the environmental movement at this time was immense and continued to build. In 1972 the Apollo 17 crew took a series of photographs of Earth from 22,000 miles away. One of these photos, NASA 22727, became an icon of the environmental movement. Using the metaphor "Spaceship Earth," this picture came to represent a fragile Earth with a delicate natural balance. Since humanity lived on this globe, its collective fate was linked to the fate of the Earth (Cosgrove 1994: 270–294).

When Richard Nixon took office as president, he announced that he was an environmentalist and that he supported the development of legislation to protect the environment. Many of the major U.S. environmental laws were passed during the Nixon administration. They include the National Environmental Policy Act (1970), the Federal Water Pollution Control Act (1972), and the Coastal Zone Management Act, the Federal Pesticide Control Act, and the Marine Mammal Protection Act (1974). In addition, the U.S.

Figure 8.2
A photograph of Earth from space taken during the Apollo 17 mission in December of 1972.

Environmental Protection Agency was created in 1970 to consolidate the federal government's efforts.

Professionalization and Insulation

The election of Ronald Reagan in 1980 brought a halt to any major initiatives in environmental legislation. Throughout the first four years of his administration, the agencies that were charged with enforcing federal environmental laws were systematically attacked, and their budgets reduced. Owing to scandals at EPA and the Department of the Interior, and the emergence of the environment as a significant issue in the 1984 elections, the Reagan administration muted its attacks on environmental agencies and laws.

In this conservative political atmosphere, a new environmentalist strategy was developed. In an editorial in the *Wall Street Journal*, Fred Krupp, executive director of the Environmental Defense Fund, maintained that the environmental movement had entered a third stage (Krupp 1986). Teddy Roosevelt and John Muir represented the first stage of the movement, according to Krupp. The efforts at this time were to halt the "truly rapacious exploitation of natural resources." The second stage started with the publication of *Silent Spring*. The aim of this phase of reform environmentalism was to work to halt abusive pollution through lobbying, lawsuits, and other direct actions. The aim of the third stage was to find alternatives to ecologically destructive projects. Behind these projects, Krupp maintained, there are "nearly always legitimate social needs." The long-term solution for environmental problems, then, is "to move beyond confrontation, and work with new coalitions to develop new ways to address persistent environmental problems, and to create environmentally-sound economic growth" (Tober 1989: 31). This approach to environmental problems led the Environmental Defense Fund and several other Reform Environmentalist organizations to develop ties to the business and industry communities. In addition, there was the development of behind the scenes negotiations between the leading Reform Environmental groups and industry representatives. This strategy is controversial. The EDF argues for this form as a way of mutually working out solutions to difficult problems, but several other environmental groups see the EDF as engaging in an illegitimate and elitist strategy. This controversy splits reform environmentalism today (Gottlieb 1993; Dowie 1995).

Current Organizations

Organizations based in the discourse of Reform Environmentalism identify their purpose as protecting the Earth's ecosystem and human health. These organizations tend to use phrases along the lines of "to protect and enhance human welfare and combat environmental deterioration" and "this organization is dedicated to improving environmental quality and public health." The magazines produced by these organizations describe ecological problems and propose ways to remedy them. In general, the descriptions of the problems are very analytic and are based in the natural sciences. The proposed remedy is usually some form of law or regulation to be enforced by government, or tax incentives to use market forces to clean up or reduce the ecological problem. In addition, the successes of various organizational initiatives are listed, generally as a regular feature of the magazine.

To examine the characteristics of organizations informed by this discourse, I compiled and examined a sample of 31 Reform Environmentalist groups. The founding dates of most of these organizations fall between the late 1960s and the early 1980s. The mean founding date is 1969. These organizations are moderate in size, with an average membership of 96,000 and an average staff of 86.

These groups have substantial financial resources, with an average annual income of $6.9 million and average assets of $4.6 million. The sources of their income are shown in table 8.1. Foundation grants are the source of more than one-third of these groups' income and the largest single source of funding.

To influence policy, 48 percent of these groups follow primarily an education strategy and 45 percent primarily a parliamentarian strategy. However, 7 percent follow primarily a protest strategy. This is due to the tactics of Greenpeace, which shares many of the goals of the other Reform Environmental groups but is more confrontational.

More than two-thirds of these environmental groups are oligarchic in structure. In the other third, limited democracy is the most common form of internal organizational structure.

Social Learning and Reform Environmentalism

Since 1970, Reform Environmentalism had replaced Conservation as the dominant discourse through which the problem of environmental degradation is viewed in U.S. society. There are numerous groups that focus on

Table 8.1
Sources of income of reform environmentalist groups.

	Percentage of income
Public contributions	34.1
Foundation grants	36.7
Government grants	4.0
Program services	8.9
Membership dues	5.4
Other	10.9

virtually every ecological problem using this discursive perspective. Thus, it is a wide and encompassing discourse that can be used to comprehensively examine virtually any ecological issue. Accordingly, it is not limited in its ability to connect with a large range of different audiences.

However, since reform environmentalism's initial successes in the 1970s, there have been almost no new major environmental initiatives. One exception to this trend was the development of the Montreal Protocol to address the ecological problem of ozone depletion. Since the 1970s, the extent and pace of ecological degradation has accelerated. New ecological problems, including the proliferation of endocrine disrupters, biodiversity loss, and global warming have been identified, without any meaningful actions being taken to address them. Schlosberg (1998: 10), commenting on current reform environmentalism, stated: "It has failed, more recently, on two key counts. The first is the failure of the major groups to continue to secure gains within the model and to satisfy even the limited interests it represents. And the second is the obvious inadequacy of these groups in representing the diversity of environmental interests, identities, foci, and forms of action." Why, despite the proliferation of environmental groups, is this the case?

The discourse of Reform Environmentalism limits the social learning capacity of our social order. Its scientific analysis of environmental problems has developed a strong critique of the ecological effects of our current institutional structure. However, this alone is not sufficient to develop an alternative vision on which an ecologically sustainable society can be based (Killingsworth and Palmer 1992: 19). Reform environmentalism is primarily based on the writings of natural and physical scientists. With the exception of Commoner, the vast majority of ecological scientists have not examined the social and political causes of ecological degradation (Taylor 1992: 133–151). While the natural scientists may have great competence in their specific areas of expertise, their social and political thinking is "marred by blindness and naiveté" (Enzensberger 1979: 389). Thus, this discourse obscures consideration of the social factors that cause ecological degradation. "The destruction of mankind," Enzensberger writes, "cannot be considered a purely natural process. But it will not be averted by the preaching of scientists, who only reveal their own helplessness and blindness the moment they overstep the narrow limits of their own special areas of competence." (ibid. 1979: 393) The problem with this form of analysis is not that it is empirically wrong, but rather that it is partial. It can identify the ecological consequences of our practices, but it fails to consider their social origins.

Reform environmentalism has been unable to develop a meaningful political vision of how to create an ecologically sustainable society. Without this vision, it lacks the means to engender widespread political support. The result is that "modern environmentalism—is politically naive and perhaps, irrelevant" (Taylor 1992: 136). Instead, environmental politics takes the form of technical and legal debates carried out within a limited community of lawyers and scientists that is heavily biased in favor of industrial interests.

In addition, reform environmentalism fosters practices that limit its political mobilization and social learning capacity. Specifically, it fosters the creation of oligarchic practices and "Astroturf" groups, thus limiting public participation in the development and advocacy of environmental policy. Reform environmentalism defines a practice of science-based piecemeal reform. The solution to environmental problems lies in top-down system management (Taylor 1992: 27–50; Evernden 1992a). Scientists, because of their expert role, are seen to need to play a prominent role in the "scientific management of the environment" (Taylor 1992: 50). Citizens and the "decision makers" play only bit parts, heeding the advice of the scientist. There is no identified need to involve the public, except to obtain financial support. The political implications of reform environmentalism have amounted to "authoritarian management of the environment and society" (ibid.: 49). Even the much-used symbol of "Spaceship Earth" has ideological functions (Cosgrove 1994). By projecting a vision of one world, it erases social and economic divisions and functions as "an insidious technique for justifying the political status quo with its accompanying inequality and injustice" (Taylor 1992: 44).

Thus, it is not surprising that the discourse of Reform Environmentalism fosters the development of oligarchic organizations. More than two-thirds of Reform Environmentalist groups have this form of structure. This high degree of oligarchic structures fosters the development of groups that have a high potential to become isolated from the members they "represent." In addition, the significant economic support these groups receive from foundations indicate that these groups have a substantial potential to be channeled and co-opted by external funding organizations. Together, these two characteristics lead to a high probability that the discourse of Reform Environmentalism will foster the development of Astroturf organizations. This compromises the independence of a movement organization, and it ties the organization back to subordination and control by the market and bureaucratic state. So instead of standing outside of these two systems of instrumental reason, Reform Environmentalist organizations have a high potential to become controlled by political and economic interests.

9

Alternative Voices

After the development of Reform Environmentalism, the discourses of Conservation, Preservation, and Wildlife Management became subsumed under the broad and diffuse label of environmentalism. Just as Conservation had achieved dominance as "the" frame that described concern about the natural environment during the period 1900–1968, Reform Environmentalism took over in 1970. As the limitations of this frame became apparent, and the progress toward resolution of ecological problems stalled in the 1980s and the 1990s, a number of alternative discourses and social movements developed. In this chapter I discuss the development of four of these alternative discourses: Deep Ecology, Environmental Justice, Ecofeminism, and Ecotheology.

Deep Ecology

All ethics so far evolved rest upon a single premise: that the individual is a member of a community of interdependent parts. His instincts prompt him to compete for his place in the community, but his ethics prompt him also to co-operate. . . . The land ethic simply enlarges the boundaries of the community to include soils, waters, plants, and animals, or collectively; the land.
—Aldo Leopold (1949: 239)

John Muir said that if it ever came to a war between the races, he would side with the bears. That day has arrived.
—Dave Foreman (1985: 11)

The first alternative discourse to appear in the 1980s was Deep Ecology.[1] At the core of this perspective is the belief that "intuition about the intrinsic value of all nature that will ground a respectful way of living in and with the natural, non-human world" (Rothenberg 1992: 50). This perspective views that a change in mindsets and worldviews will result in a change in the social order (Oelschlaeger 1991: 301–309 Merchant 1992: 85–109). In Deep Ecology, nature is seen as a value in its own right, independent of human existence. Humanity is only one species among many, and has no right to dominate the Earth and all the other living organisms. This creates an ethic of radical wilderness advocacy. Unlike Preservation, which seeks to keep what remains, Deep Ecology seeks the restoration of fully functioning ecosystems in which the evolution of life, unaffected by human actions, can continue. It also advocates the inherent rights of all nonhuman beings to exist in their natural state. In this sense, Deep Ecology makes a moral argument for the preservation of the natural environment. The following are key components of this discourse (Devall and Sessions 1985):

- The richness and diversity of all life on Earth has intrinsic value.
- Humankind's relations to the natural world presently endanger the richness and diversity of life.
- Human life is privileged only to the extent of satisfying essential needs.
- Maintenance of the diversity of life on Earth mandates a decrease in the human impacts on the natural environment, and substantial increases in the wilderness areas of the globe.
- Changes (consistent with cultural diversity) affecting basic economic, technological, and cultural aspects of society are therefore necessary.

Development of the Idea of Deep Ecology
The writings of the zoologist Nathaniel Shaler foreshadowed the development of this discourse. These early works discussed the need for the development of a new ethical relationship with the natural world that would protect the Earth and all its creatures, including man. "We are parts in a living sensitive creation," Bailey (1915: 30) argued. "The living creation is not exclusively man-centered; it is biocentric. . . . We can claim no gross superiority and no isolated self-importance. The creation, and not man, is the norm."

This work had a direct influence on the thought of Aldo Leopold. In his classic essay "The Land Ethic" (1949), Leopold argues for the extension of ethics to include the relationship between humans and the natural environment: "The land ethic simply enlarges the boundaries of the community to include soils, waters, plants, and animals, or collectively; the land." Leopold's writings have served to further inspire the development of a discourse that centers on the value of all living things.

The term Deep Ecology emerged in 1972 in the writings of the Norwegian philosopher Arnie Naess (Devall 1985: 65, 225–228; Borecelli 1988: 32). In a 1972 article titled The Shallow and the Deep, Long Range Ecology Movement, A Summary, Naess defined Shallow Ecology (describing it as a "fight against pollution and resource depletion") and Deep Ecology (which he saw as a deeper concern of ecology with the structure and purposes of society, "which touch upon principles of diversity, complexity, autonomy, decentralization, symbiosis, egalitarianism, and classlessness"). The focus of the distinction between shallow and deep ecology is whether the natural world is a resource for human use or whether nature has a right to exist regardless of its utility for human purposes.

Also highly influential in the development of Deep Ecology was the poet Gary Snyder, whose 1974 collection *Turtle Island* extended Leopold's ideas into a more comprehensive view of the proper relationship between humanity and nature. In *Turtle Island*, Gary Snyder defines an alternative viewpoint through poetry (Oelschlaeger 1991: 243–280). One poem, "Tomorrow's Song," reads in part as follows:

The USA slowly lost its mandate in the middle and later twentieth century
it never gave the mountains and rivers, trees and animals, a vote.
all the people turned away from it
myths die; even continents are impermanent
Turtle Island returned. . . .
We look to the future with pleasure
we need no fossil fuel
get power within
grow strong on less

Beginnings of Collective Action

Deep ecology was used as the basis for collective action in the early 1980s. The inspiration for the formation of a Deep Ecology group was found in Edward Abbey's 1975 book *The Monkey Wrench Gang*. This movement grew out of a rejection of Reform Environmentalism. Part of this rejection was based on the bureaucratization and professionalization of the Environmental Movement (Manes 1990: 45–65). One of the founders of Earth First! explained it as follows: "Mainstream environmentalists are out of touch. . . . Most of them are in D.C. doing lunch in their designer khakis and working out their retirement bennies. The problem is, the environment isn't a calling anymore, it's a job." (Cawley 1993: 160)

Earth First! was founded in 1980. This action marked the development of the ideas of Deep Ecology into collective action (Scarce 1990: 57; Manes 1990: 1–7). One of Earth First's founders wrote that it had been founded "to state honestly the views held by many conservationists, to demonstrate that the Sierra Club and its allies were raging moderates, to balance . . . anti-environmental radicals, to return vigor, joy and enthusiasm to the tired, unimaginative environmental movement, to keep the established groups honest, to give an outlet to many hard-line conservationists, to provide a productive fringe, to inspire others to carry out activities straight from the pages of *The Monkey Wrench Gang*, and to help develop a new worldview, a biocentric paradigm, an Earth philosophy. To fight, with uncompromising passion, for Earth." (Foreman 1991: 18)

Earth First! is a self-described adherent to the philosophy of Deep Ecology. A 1987 information handout asserts two key aspects of the group's beliefs. The first is the idea of the interconnectedness of all life. The handout quotes Aldo Leopold: "A thing is right when it tends to preserve the integrity, stability and beauty of the biotic community. It is wrong when it tends otherwise." The second point is that "all natural things have intrinsic value, inherent worth"—that the value of natural things "is not determined by what they will ring up on the cash register of the GNP, nor by whether they are good. . . . They are. They exist for their own sake. Without consideration for any real of imagined value to human civilization. Accordingly, Earth First! maintains its purpose to be defenders of the Earth. Wilderness, natural diversity, is not something that can be compromised in the political arena. We are unapologetic advocates for the natural world, for Earth."   The first action of Earth First! focused on the Glen Canyon Dam. This dam has special significance to this group and other wilderness advocates. It was built in the late 1950s on the Colorado

River, filling up Glen Canyon. This canyon had many Indian artifacts and dwellings in it. In addition, its beauty was said to rival the Grand Canyon. The beauty of this area was captured in a series of photographs that appeared in the November 1958 *Sierra Club Bulletin* and in a retrospective book on the canyon, *The Place No One Knew* (Porter and Brower 1966). To dam up such a place is considered a sacrilege by many environmentalists. This dam became a natural target for environmental-movement groups seeking to denounce development (Martin 1989).

In 1981 a group of Earth First! activists decided to make a powerful symbolic statement using the Glen Canyon Dam as the focal point. Accompanied by Edward Abbey, Earth First! activists hung a large plastic crack on the dam. The importance of this action was that, for the first time, an environmental group in the United States had called for the reversal of a major development project. Though fighting dams has been a tradition of America environmentalism since the Hetch Hetchy battle in the period 1910–1915, no group had previously called for tearing down a dam and restoring a river valley to its natural state.

Another substantial movement based on the discourse of Deep Ecology is bioregionalism, which defines an alternative framework on which to organize human communities. In contrast to political boundaries, bioregionalism advocates organizing human communities in relation to a specific ecological place or bioregion. A bioregion, as define by Peter Berg (Merchant 1991: 218), is a "geographic area having common characteristics of soil, watersheds, climate, and native plants and animals." This type of community organization would re-frame human existence as part of a natural ecosystem, not apart from it.

Planet Drum Society is the leading U.S. environmental organization advocating bioregionalism (Merchant 1991: 217–222; Borecelli 1988: 33–35; Parking 1989: 297). An attachment to IRS Form 1023 dated September 14, 1977, states the purpose the Planet Drum Society as follows:

Planet Drum was founded on the premise that humans share space on the planet with other species of animal and plant life, and that we need to learn to live in greater harmony with these non-human inhabitants if we want to insure long-range abundance and prosperity. Planet Drum's purposes are to do research about the relationships between humans and other species and then to educate the public about ecologically positive ways of living.

A number of bioregional settlements associated with Planet Drum define themselves by unique names that correspond to their ecological bioregions.

Biodiversity and the Society of Conservation Biologists

Deep Ecology also inspired the formation of the academic discipline of conservation biology. The founders of this academic discipline see it as a unification of ecology and evolutionary biology, with a normative orientation to conserve biological diversity. Their mission is "not merely to document the deterioration of Earth's diversity but to develop and promote the tools that would reverse that deterioration" (Takacs 1996: 35).

Conservation biology emerged into national view in September of 1986 at the National Forum on BioDiversity. Held in Washington under the sponsorship of the National Academy of Science and the Smithsonian Institution, this conference was attended by 14,000 people. A group of eminent biologists redefined and publicized the problem of endangered species. The conference "signaled a new direction in conservation thinking and action" (Takacs 1996: 39). At the heart of this change in thinking was the development of the idea of *biodiversity*. The National Forum on BioDiversity legitimized this idea. Although this term only came into existence in the mid 1980s (Takacs 1996: 37), by the mid 1990s it had already become a key concept around which the debate regarding the proper relation of human society to the natural environment revolves.

The term *biodiversity* was developed by conservation biologists to spur political action. These biologists "believe that humans and the other species with which we share the Earth are imperiled by an unparalleled ecological crisis," and "biodiversity is the rallying cry currently used by biologists to draw attention to this crisis and to encapsulate the Earth's myriad species and biological processes, as well as a host of values ascribed to the natural world" (Takacs 1996: 9). In this context, it was argued, the term can best be described by means of ideograph—"a high-order abstraction representing collective commitment to a particular but equivocal and ill-defined normative goal" (McGee 1980: 15).

The leading organization promoting this unique form of scientific activism, or normative science, is the Society for Conservation Biology, the purpose of which is "to help develop the scientific and technical means for the protection, maintenance, and restoration of life on this planet—its species, its ecological and evolutionary processes, and its particular and total environment" (http://conbio.rice.edu).

Conservation biologists have begun working with deep ecologists, such as Dave Foreman, to develop projects to enhance biodiversity. One such project is the Wild Lands Project, whose mission is "to help protect and restore the ecological rightness and native biodiversity of North America through the establishment of a connected systems of reserves" (Wild Earth 1992: 3).

In addition, conservation biologists have formed alliances with financiers, including the multimillionaire Douglas Tompkin. One result of the alliance with Tompkin was the founding, in 1989, of the Foundation for Deep Ecology, which funds activities in support of rainforest preservation, grass-roots activism, supporting indigenous peoples to retain their sovereignty, and opposing economic changes that adversely impact Third World peoples and the natural environment. This foundation's assets exceed $35 million. Its board members include Dave Foreman and several conservation biologists (Foundation Center 1997).

One legislative proposal to enact the idea of biodiversity into a practical application is the Northern Rockies Ecosystem Protection Act (HR 95-852), introduced on February 7, 1995. Its aim is to ensure the protection of all the major ecological systems and species in the Northern Rockies. To accomplish this task, the bill provides for adequate acreage for the national parks in the area of concern, and for the creation of corridors between the parks to maintain their ecological integrity.

Another ongoing effort is the development of a formal organization to advocate removal of the Glen Canyon Dam. The Glen Canyon Institute was founded in 1996 with the mission of providing "leadership in reestablishing the free flow of the Colorado River through a restored Glen Canyon" (Glen Canyon Institute 1996). Hearings were held in the U.S. House of Representatives on September 23, 1997, to debate this proposal. David Brower and other members of the Glen Canyon Institute appeared as an advocates for draining the reservoir. The hearings were planned by western lawmakers to embarrass the proponents of restoring Glen Canyon. However, the hearings worked against the desires of the lawmakers. Not only did the proposal to drain Lake Powell make economic and ecological sense; in addition, the fact that the hearings were held at all legitimized the idea (*New York Times*, October 6, 1997).

The Zero Cut Campaign and the Reinvigoration of the Sierra Club

The idea for the reinvigoration of the Sierra Club emerged in the early 1990s among a group of forest activists. The idea was that if the environmental movements were ever to be able to make a real impact on changing existing practices, the Sierra Club would have to increase its activism on behalf of wilderness protection. In a 1991 article in *Wild Earth*, the environmental activist Hart Schaefer argued that "we [forest activists] need the Sierra Club if we are to save anything. We need the Sierra Club to be what it claims to be, what most people still think that it is." The article goes on to state the Sierra Club's rules about referendums and club elections, and suggests that activists make a determined effort to use these measures to reinvigorate the club.

To enact these ideas, a group of activists led by David Orr formed a loose coalition of approximately 50 or 100 like-minded individuals in the Sierra Club (Dowie 1995: 216–217). (They are now known as the John Muir Sierrans.) This effort, paralleling the efforts of David Brower in the 1950s, is now conducting a campaign to revitalize the Sierra Club and to make it a leader in protecting the natural environment.[2] In April of 1996, these activists succeeded in placing a referendum question on the Sierra Club annual election ballot. The question, now known as the Zero Cut Initiative, read as follows: "Shall the Sierra Club support protecting all federal publicly owned lands in the United States by advocating an end to all commercial logging on these lands?" (Sierra Club Ballot, 1996) After a bitter debate within the club, the proposal succeeded, getting more than 60,000 votes out of about 90,000. In the same election, some John Muir activists were elected to the national board of the Sierra Club. In 1998, three more John Muir Sierrans were elected to the national board, thus giving this reform effort a powerful influence in determining Sierra Club policy.

Whether this effort will serve as a basis for further revitalization of the entire environmental movement, as David Brower's efforts did in the 1950s, is yet to be determined. However, to date, this effort has had significant impacts on forestry policy. Advocacy of the Zero Cut Initiative by the Sierra Club and a number of other forest-protection groups resulted in introduction of the National Forest Protection and Restoration Act (HR 2789) on October 31, 1997. This bill's purpose is to reverse the policy enacted in the Organic Act of 1897, which allowed commercial logging in the National Forests.

Current Deep Ecology Organizations

Organizations based in Deep Ecology generally define their objectives as acting to preserve the rights of all nonhuman beings to a natural existence unaffected by human intervention. These organizations use words such as "intrinsic rights of species to life" or "placing ecological considerations first in any decision making process" to define their purpose. The publications of these types of organizations are almost always printed on newsprint. The articles in these publications focus on assaults on living species, and interventions by the organization in their behalf. Using headlines such as "Dolphins Kidnapped" to describe an aquarium's capture of dolphins for exhibit, these magazines describe the destruction of wilderness and wildlife as assaults on the inherent rights of these beings. In addition, since many Deep Ecology magazines engage in direct action to realize their goals, the magazines detail the exploits of the members of the organization as heroes protecting all living beings.

These organizations issue a large variety of graphical material that can be ordered from their magazines.[3] There are small "Silent Agitator" stickers, designed to be placed in areas such as the rest rooms of government agencies or large corporations to remind the individuals who work there of opposition to their policies. There are T shirts with various Earth First! symbols on the front, bumper stickers and signs with messages such as "Developers Go Build in Hell" and "Subvert the Dominant Paradigm," music with titles such as "The Ballad of the Lonely Tree Spiker," and books such as *Ecodefense: A Field Guide to Monkeywrenching* (Foreman and Haywood 1985). One of the most interesting and representative illustrations appears on the first page of *Ecodefense*: a heroic monkeywrencher, armed with a monkey wrench and a hammer and garbed in a superhero uniform emblazoned with the Cobb symbol,[4] strides through a break in a barbed wire fence.

Figure 9.1
Ecowarrior (from the inside cover of Foreman and Haywood 1985).

To examine the characteristics of organizations informed by this discourse, I compiled and examined a sample of eight Deep Ecology groups. The founding dates of most of these organizations fall between the late 1970s and the early 1990s. The average founding date is 1981. These organizations are very small in comparison to other environmental-movement groups, with an average membership of 12,000 and an average staff of 10.

These groups have few financial resources, with an average annual income of just over $500,000 and total assets of just above $10,000. Their sources of income are shown in table 9.1. Some Deep Environmental groups derive a large portion of their income from foundation support. As table 9.1 shows, foundation grants, primarily from the Foundation for Deep Ecology, constitute the second largest source of income for these groups, accounting for nearly one-third of their income.

Table 9.1
Sources of income of deep ecology groups.

	Percentage of income
Public contributions	39.0
Foundation grants	30.7
Government grants	0.4
Program services	10.9
Membership dues	6.1
Other	12.9

These are the most diverse of the environmental organizations in terms of their strategies. Every strategy for influencing policy has been used by at least one group. However, fully 50 percent of these groups engage in a protest strategy.

In organizational structure, 100 percent of these groups are oligarchic. (Earth First! is omitted from this analysis because it has no formal organizational structure.)

Social Learning and Deep Ecology

In examining the discourse, Taylor (1992: 97) maintains that "the two ultimate norms of deep ecology . . . are the promotion of 'self realization' and 'biocentric equality'" and that "self-realization is to be achieved by widening our understanding of self to include the natural world—apparently a twofold process of overcoming a narrow egoistic understanding of self-interest, and simultaneously developing a sympathy with other living things." It is clear from this analysis that Deep Ecology is a limited discourse that focuses on the protection of wilderness and wildlife. In addition, the focus on the transformation of self and on the development of a new relationship with the natural world reduces its relevance for the analysis of political and social questions. Deep ecology focuses on preserving and restoring the natural world through extension of self-understanding and development of personal sympathy with other living things. The manifestations of Deep Ecology are primarily in the form of aesthetic expressions of this alternative relationship with nature. This aesthetic perspective opens up the space for consideration of alternative value systems and ways of being in the world (Killingsworth and Palmer 1992: 20). At the same time, however, this focus leads to a neglect of the political and social changes necessary to inform the creation of an ecologically sustainable society. This lack of a political dimension to this thought "prevents Deep Ecologists from being able to describe and defend a radical alternative social and political life informed and guided by their own environmental ethics" (Taylor 1992: 131). Instead, the political actions of Deep Ecology focus almost exclusively on the defense of wilderness, with virtually no efforts being extended on the reform of society. The extension of Deep Ecology to human concerns vacillates "between overt hostility toward the human community in general and a vague appeal to the extension of the human community to the broader natural world" (Taylor 1992: 131–132). This inability to address political questions leaves questions about the means of social transition to this alternative vision unanswered. This limits its ability to inform alternative political and social practices.

The extraordinary incidence of oligarchic structures among Deep Ecology groups is probably due to a combination of two factors. The first factor is a perceived need on the part of radical groups to centralize power in the face of a hostile social environment. The second (and more troubling) factor is that oligarchic practices seem to be enabled by the discourse of Deep Ecology. As was discussed above, Deep Ecology is primarily based on individual insights and on the development of a "new" self-understanding. This could create divisions between those who have gained access to this "new" form of self-understanding and those who haven't. The new self-understanding and insight could take on a form of revealed truth, justifying control of these organizations by those individuals who have achieved access to this truth. Such organizations would then tend to be controlled by charismatic leaders whose power stems not from democratic procedures but from their access to unique insights.

In addition, Deep Ecology has no means of addressing social and political considerations. This discursive blindness could result in the adoption of the most expedient means to achieve the protection of wilderness, without regard for social and political consequences. In addition, the high level of foundation support indicates that these groups have a substantial potential to be channeled into certain areas by external funding organizations.

Together, these characteristics lead to a high potential for Deep Ecology organizations to become isolated from their members over time.

Environmental Justice

The City of Niagara Falls, the State of New York, the federal government, the EPA, the mayor . . . all knew that those poisons were in my backyard. . . . And they made a decision, a decision that is being made in each and everyone of your communities. They decided, because my husband made $10,000 a year, because we were working class people, that it was OK to kill us. That was the decision that they made, and it was that terrible, awful, tragic experience that made me realize that there is nobody out there who is going to look out for Lois Gibbs and her neighbors—except for Lois Gibbs and her neighbors."
—Lois Gibbs, 1993

There were mass movements for social justice . . . to end slavery and for women's suffrage. But now all of those rights are threatened by . . . the abuse of the planet in which those rights might be exercised or implemented. And thus we must see in our fervor for rights that without the right to breath, nothing else really matters.
—Rev. Jesse Jackson (quoted in Lapp 1992)

The second alternative voice to emerge in the 1980s was the discourse of Environmental Justice, which sees the source of ecological problems in the structure of our society (Oelschlaeger 1991: 307–308; Merchant 1992: 132–154; Borecelli 1988: 38; Sandbach 1980: 26–27). In this discourse, the link between human survival and ecosystem survival defined by Environmentalism is accepted. However, instead of focusing on the physical causes of environmental degradation, this frame sees environmental problems as creations of human social order. Hence, the solution of environmental problems lies in social change.[5] The following are key components of this viewpoint:

• Domination of humans by other humans leads to domination of nature.
• The economic system and nation-state are the core structures of society that create ecological problems.
• Commoditization and market imperatives force consumption to continually increase in the developed economy.
• Environmental destruction in low-income and racially distinct communities or in Third World countries originates in the exploitation of the people who live in these areas by the dominant social institutions.

• Resolution of environmental problems requires fundamental social change based on empowerment of local communities.

Organizations based on the discourse of Environmental Justice focus on the social creation and resolution of environmental problems (Capek 1993). Taylor (1993: 57) summarizes the key aspects of the movement as follows:

It integrates both social and ecological concerns much more readily and pays particular attention to questions of distributive justice, community empowerment, and democratic accountability. It does not treat the problem of oppression and social exploitation as separable from the rape and exploitation of the natural world. Instead, it argues that human societies and the natural environment are intricately linked and that the health of one depends on the health of the other.

In addition, the discourse of Environmental Justice sees the use of scientific experts as part of a system of oppression and domination. Without

Figure 9.2
"Experts and the public" (from *Experts, A Users Guide,* published by Citizens' Clearinghouse for Hazardous Waste).

access to experts of their own, some local community activists see scientific discussions as a means of keeping their viewpoints and concerns from being addressed by government officials. As a consequence, environmental justice groups challenge the authority of scientific experts to adequately express community concerns.

Development of the Discourse of Environmental Justice

Concern about the inequitable impact of environmental problems can be traced back to Chadwick's 1842 report, which viewed disease as attributable to poverty and poor sanitation. Thus, the living conditions imposed by economic conditions on the poor were seen as harmful to their health. Urban environmental conditions and the poor were also concerns of the urban reformers of the early nineteenth century, of the sanitary movement, and of the urban housekeeping movement. However, the discourse of Environmental Justice did not develop until the 1980s.

The development of the idea of environmental justice springs from two related systems of ideas dealing with social institutions and ecological degradation. The ideas are based on the idea that class position in a capitalist society and on the idea that racism systematically results in a disproportionate share of ecological degradation being borne by the poor and ethnic minorities.

As academic scholars were developing their arguments and concepts, activists were demanding a livable environment. In addition, there were several activist-scholars who worked on the analyses for local communities in addition to publishing academic work. These two groups learned from one another. Although the development of the frame of environmental justice is presented in two parts, it is important to keep in mind this co-development by academics and activists.

Capitalism and Environmental Degradation

A discussion of the relationship between capitalism and environmental degradation appeared in Petr Alekseevic Kropotkin's analysis of the social origins of ecological problems (1899/1994). Kropotkin maintained "that industrialism is unsustainable because of its excessive strains on the natural environment" (Roussopoulos 1993: 16). Barry Weisberg further expanded this idea in his 1971 book *Beyond Repair*, arguing that the inhuman and anti-ecological bias of the capitalist system of production was the source of ecological imbalance. This analysis was further elaborated by a number of studies examining the origin of environmental problems in the political economy of advanced capitalist economies (Schnaiberg 1980; Schnaiberg and Gould 1994; Cotgrove and Duff 1980; Cotgrove 1982; O'Connor 1973, 1984, 1987). Originating in neo-Marxist economic analysis, these approaches assign primacy for the origin of environmental problems to the logic of the mode of production. One major work in this area is Alan Schnaiberg's *The Environment: From Surplus to Scarcity* (1980). As I noted in chapter 3, Schnaiberg argues that the capitalist economy forms a "treadmill of production" that continues to create ecological problems.

Perhaps the most comprehensive theoretical statement on the unequal distribution of environmental risk is contained in Ulrich Beck's book *Risk Society* (1986). Beck defines the idea of risk positions that characterize the levels and the nature of technological risk to which people are exposed (p. 23). He goes on to characterize the distribution of risk positions as follows (p. 35):

> The history of risk distribution shows that, like wealth, risks adhere to the class pattern, only inversely; wealth accumulates at the top, risks at the bottom. . . . It is especially the cheaper residential areas for low-income groups near centers of industrial production that are permanently exposed to various pollutants in the air, the water and the soil. . . . Here it is not just this social filtering or amplification effect which produces class specific afflictions. The possibilities and abilities to deal with risk, avoid them or compensate for them are probably unequally divided among the various occupational and educational strata.

Environmental Degradation, Class, and Race

Empirical analysis of the relationship between different risk positions and social strata pointed to by Beck originated in the 1970s. One early study to document the relationship between socio-economic position and levels of exposure to pollution appeared in *Science* in 1970 (Lave and Seskin 1970: 728). In an examination of 114 metropolitan areas, race and economic position were found to correlate with increased exposure to air pollution. This study was expanded in an examination of exposure to air pollution in New Haven by William Burch, who examined the relationship between air quality and both race and class and noted that "class rather than race may be the more salient factor in determining exposure to the health hazards of air pollution" (Burch 1976: 313).

The relationship between socio-economic conditions and environmental degradation was also the subject of a chapter in the Council on Environmental Quality's Second Annual Report (CEQ 1971: 189–207), which examined a number of urban environmental problems—including air pollution, water pollution, solid waste, neighborhood deterioration, lead poisoning, access to open space, opportunities for recreation, and transportation policies—and concluded with a call for joint action by citizens and government.

These early findings were amplified by a series of studies pertaining to the siting of hazardous-waste dumps. A 1983 study conducted by the General Accounting Office documented that African-American communities had a higher than average siting rate of waste dumps (U.S. GAO 1983). A 1987 report by the United Church of Christ titled Toxic Waste and Race in the United States documented the unequal and discriminatory siting of toxic-waste facilities in the United States (UCC 1987). In the latter report, the executive director of the United Church of Christ Commission for Racial Justice, Rev. Benjamin Chavis, defined "environmental racism" as "racial discrimination in environmental policymaking, the enforcement of regulations and laws, the deliberate targeting of communities of color for toxic-waste facilities, the official sanctioning of the life-threatening presence of poisons and pollutants in our communities, and the history of excluding people of color from leadership of the ecology movements" (Bullard 1990: 278).

In 1987 the empirical research about the unequal and disproportionate siting of toxic-waste facilities in poor and minority communities was lent additional weight by the unauthorized release of a report titled Political Difficulties Facing Waste-to-Energy Conversion Plant Siting (Cerrell Associates 1984), which had been prepared under contract for the California Waste Management Board. In its third chapter, the report examined the characteristics of communities where local political resistance to the siting of municipal waste and toxic-waste incinerators would be the least. It suggested that "people who are resentful and people who are amenable to Waste-to-Energy projects could be identified before selecting a site," and that "if this information was available, facilities could be placed in areas, if technically feasible, where people do not find them so offensive" (ibid.: 17). The study went on to identify the characteristics of these communities. The key characteristics of less resistant communities were identified as low income, low education, high minority population, and openness to promises of economic benefits from the proposed facility.

In 1990, a group of scholars and activists organized a Conference on Race and the Incidence of Environmental Hazards at the University of Michigan. The results of this conference, published in 1992, clearly documented and "overwhelmingly corroborated the evidence of the General Accounting Office and the United Church of Christ reports" (Bryant and Mohai 1992: 3). Robert Bullard's 1990 book *Dumping in Dixie* compiled the evidence to date and firmly documented the existence and the effects of environmental racism.

Since 1990, an extensive and elaborate literature on differential environmental risk based on class position and racial community has developed. These studies, as summarized by Bunyan Bryant (1995), show that poor and minority communities suffer a disproportionate share of environmental degradation and exposure to toxic waste and pollution, that the populations of these communities have higher incidence of toxic induced or aggravated diseases, and that enforcement or remedial actions by government are less timely and less effective than similar actions in other types of communities.

In 1994, building on his earlier work, Robert Bullard published *Unequal Protection*, in which he argued that "the goal of an environmental justice framework is to make environmental protection more democratic." "More important," Bullard continued, "it brings to the surface the ethical and political questions of who gets what, why, and in what amount." Bullard (ibid.: 10–11) defined the basic elements of the framework of environmental justice as follows:

1. Incorporates the principle of the right of all individuals to be protected from environmental degradation,
2. Adopts a public health model of prevention (elimination of the threat before harm occurs) as the preferred strategy,
3. Shifts the burden of proof to polluters and dischargers who do harm or discriminate or who do not give equal protection to racial and ethnic minorities and other "protected" classes,
4. Allows disparate impact and statistical weight, as opposed to "intent" to infer discrimination,
5. Redresses disproportionate risk burdens through targeted action and resources.

The Movements for Environmental Justice

Well into the 1970s, with a few variations, the various environmental organizations in the United States reflected the interests and values of upper-class and middle-class white American males. As Jordan and Snow (1992) suggested, the class and ethnic orientation of these groups was an outgrowth of the society-wide exclusion of ethnic minorities, women, and the working class from participation in the political system.

In the late 1970s, local environmental justice groups began to form throughout the United States in working-class or ethnic neighborhoods that were experiencing high levels of environmental degradation, primarily in the form of toxic-waste pollution (Freudenberg and Steinsapir 1992). These organizations emerged from grassroots organizing activities. "In many instances," Bullard (1993: 8) writes, "grassroots leaders emerged from groups of concerned citizens (many of them women) who see their families, homes, and communities threatened by some type of polluting industry or government policy." In the white working-class community this came to be called the Citizen-Worker Movement; in ethnic minority neighborhoods it came to be called the People of Color Environmental Movement.

The Citizen-Worker Movement

The Citizen-Worker Movement is primarily composed of white working and middle-class individuals (Gould et al. 1996: 2). This movement originated in local "not in my back yard" organizations that focused on specific issues. The organizing activities in the community that lived on the former toxic-waste dump site known as Love Canal in New York State, led by Lois Gibbs, are emblematic of the development of such groups.[6]

This movement expanded quickly, widening its focus to include a variety of issues concerned with social justice. Andrew Szasz (1994) argues this shift was the result of a rejection of reform environmentalism, which led to the development of broader and more radical approaches. "Having made that 'choice,'" Szasz notes (ibid.: 151), "the toxics movement found itself squarely within a broader left/opposition culture consisting of numerous other social movements that had similar views about the root causes of various oppressions and, also, similar tactics."

One of the first national organizations based in this discourse was the Citizens' Clearinghouse for Hazardous Waste. Building on the work of the

Love Canal Homeowners' Association, the Citizens' Clearinghouse for Hazardous Waste was founded in 1981 by Lois Gibbs (Shabecoff 1993: 231–250). Following the organizational strategies of ACORN and Fair Share, this organization serves as an organizing center for a variety of local environmental justice organizations around the United States. Its purpose, according to its by-laws, is to facilitate the development of local organizations by "supporting, encouraging and providing assistance to members and citizens generally who would seek to initiate, develop and conduct programs within and/or increase public and professional awareness and understanding of the problems faced within the fields of hazardous waste, toxic chemicals, and related environmental concerns."

The Citizens' Clearinghouse for Hazardous Waste was one of the first organizations to use the term *environmental justice*. The very first use of this term may have occurred in the name of an award that the CCHW sponsored at the First National Grassroots Convention on Toxics (CCHW 1986: 4). The next such conference, held in 1989, was titled People United for Environmental Justice.

The People of Color Environmental Movement

The movement to protest the distribution of environmental risk based on class was paralleled by a movement protesting the distribution of environmental risk based on race. Though this movement shares a great deal with the Citizen-Worker Movement, its has developed its own unique identity and organizations.

Unlike previous environmental movements, the People of Color Environmental Movement was not based just in the formation of new groups. While there has been an expansion of new groups based on this discourse, a significant component of this movement involved the reformulation of the goals of existing civil rights and community organizations to include environmental concerns (Taylor 1993; Bullard and Wright 1992: 47).

Up until the 1980s, the memberships of most U.S. environmental groups were predominantly white. Blacks and other minorities were excluded, either indirectly or specifically. For example, the 1915 by-laws of the Fraternal Order of American Sportsmen limits membership to "white male citizens of the U.S." (Fraternal Order of American Sportsmen 1915: 7). With the growth of the civil rights movement, the integration of minority and working-class individuals into environmental organizations became an issue. David Brower's attempt to integrate the membership of the Sierra Club as early as 1959 was blocked by the club's board (Fox 1981: 349). In addition, there were several attempts to combine ecological concerns with social-justice concerns in the early 1970s.

One of the first organizations to advocate addressing the issue of environmental justice was Environmental Action, which originated as an ad hoc network that organized the first Earth Day. This group's primary goals and tactics were framed by an environmentalist discourse, with water and air pollution forming its central issues. However, its founding principles also showed a concern with issues that would later be defined as "environmental justice" and "environmental racism":

If the ecology movements is to succeed, it must not set itself up as something completely distinct and apart from the other driving concerns of our times. . . . The environment is relevant to Black people and poor people, for example, because they face the worst environmental conditions in the country. Their sewage is the first to stop up, their houses are located downwind from the factory, their neighborhoods are displaced by freeways. . . . The environment is relevant to workers. Frequently the stench in our communities is but a diluted version of the stench in our factories.

A 1972 report issued by Resources for the Future recognized that the poor had a limited role in creating ecological degradation but suffered a much greater burden: "Let us recognize that of all segments of society the poor are the least responsible for pollution in the sense that materially they consume the least. Conversely, the poor are the chief sufferers from pollution in the sense that they have the least means of insulating themselves from pollution effects."

However, in the late 1960s and the early 1970s the black community was not generally supportive of the environmental movement. In 1970 the mayor of Gary, Indiana, stated that "the nation's concern with the environment" had "done what George Wallace was unable to do: distract the nation from the human problems of Black and Brown Americans" (quoted in Ostheimer and Ritt 1976: 6). The president of the National Urban League, Whitney Young, argued for delaying dealing with environmental issues until the war on poverty was won: "The war on pollution is one that should be waged after the war on poverty is won. Common sense calls for reasonable national priorities and not for inventing new causes whose main appeal seems to be in their potential for copping out and ignoring the most dangerous and most pressing of our problems." (ibid.).

Others in the black community developed unique perspectives on environmental problems. A powerful and eloquent statement of the environmental problems of the black community—a statement that foreshadowed the development of the discourse of Environmental Racism—was made in 1970 by Nathan Hare in an article titled "Black Ecology." The environmental problems of the black community, Hare argued, were both different from and more severe than those experienced by the white community: "The environmental crisis of whites (in both its physical and social aspects) already pales in comparison to that of blacks." Hare argued that this extra environmental burden was due to the unequal social, political, and economic power of the black community, which was dominated and oppressed by a racist society.

The relationship between environmental degradation and inequity was also addressed in the early 1970s. In November of 1972 the Conservation Foundation sponsored a Conference on Environmental Quality and Social Justice in Woodstock, Illinois. The purpose of this conference was "to begin an intensive dialogue on how and where the views of persons seeking environmental quality as a primary goal coincide and conflict with the views of persons seeking social justice as a primary goal" (Smith 1974: vii). This conference summarized and attempted to deal with criticisms of the environmental movement as elitist and as using environmental issues as a means to preserve its position of relative privilege. One such criticism was sounded by Norman Faramelli of the Boston Industrial Mission: "To the poor and low-income families, ecology may appear to be a cop-out, a flight from social realities, and a digression from dealing with the real issues of racism and social injustice." (ibid.: 2) In response, environmentalists argued that "if we work to achieve clean air or preserve open space it is a benefit to all, regardless of income or color" (ibid.).

The Woodstock conference successfully aired the differences between the environmental and social-justice movements. Participants saw this conference as the first step in a dialogue that might lead to the development of a unified movement working for the public interest. However, a unified movement failed to materialize at that time. Although a joint conference titled

City Care was held by the National Urban League and the Sierra Club in 1979 (Ferris and Hahn-Baker 1995: 68), there was almost no systematic contact between the environmental movement and the civil rights movement. The development of a significant movement based on this discourse did not occur until almost ten years later.

The split between blacks' and whites' environmental concerns was the subject of a 1976 study that drew the following conclusion: "Environmental problems are different for blacks: they are reflections of the socioeconomic deprivation caused by inequalities in housing, education, health care, and other basic amenities. The idea of environment exists in the contemporary thinking of black Americans within a context of social, economic, and sometimes physical oppression." (Ostheimer and Ritt 1976, p. 10)

Thus, to the black community, environmentalism as defined by the white community was based on concerns over wilderness preservation, game protection, or suburban sprawl. This definition appeared to be reflecting the concerns of the white community, and the environmental movement was seen as failing to address urban ecological problems of direct concern to the black community.

A 1990 letter written by a number of civil rights and environmental justice organizations to the leading national environmental organizations echoed the complaints about the environmental movement that had surfaced in the late 1960s: that it was nonrepresentative and fundamentally racist. This letter was followed up by demands that the existing environmental organizations hire more people of color as staffers, and that they change their policy focus to include issues of concern regarding minority communities or else stop fundraising and operations in minority communities.

Bullard (1994b: 4) traces the beginning of the People of Color Environmental Movement to a 1979 lawsuit charging racial discrimination in the location of a toxic-waste landfill in Houston. A 1982 protest against a toxic landfill in North Carolina was based in the local black community and was led and organized by a local church (Lavelle and Coyle 1993: 139; Lee 1993: 43; Bullard 1994b: 5–6). This movement was given additional impetus by the publication of the above-discussed 1983 GAO study and by the United Church of Christ's 1987 studies on environmental racism. In 1989, the Greater Louisiana Toxics March, a nationally recognized event, brought widespread recognition of the effects of the petrochemical industry on the people of southern Louisiana (Ferris and Hahn-Baker 1995: 69).

The People of Color Environmental Movement emerged as a national collective actor at a meeting in October of 1991 in Washington (Miller 1993; Grossman 1994). The Principles of Environmental Justice, a document written at that meeting, stated the movement's intention to "build a national and international movement of all peoples of color to fight the destruction and taking of our lands and communities" and listed seventeen Principles of Environmental Justice (FNPCELS 1991). This document asserted "the right to be free from ecological destruction," called for cessation of the production of all toxins, hazardous wastes, and radioactive materials and enforcement of strict accountability of the producers of these materials, and asserted "the right of political, economic, cultural and environmental self-determination of all peoples." After this conference, a number of regional networks were formed to provide aid to local groups for grassroots political activity.[7]

A significant component of this movement is the development among Native Americans of a number of organizations based on concerns about environmental protection, cultural preservation, religious rights, and enforcement of treaty obligations (LaDuke 1994; Harjo 1992). These organizations have been involved in environmental issues such as fishing rights in the Northwest and Upper Wisconsin, the effects of radiation on Navaho uranium miners, and hydroelectric development throughout North America (Whaley and Bresette 1994).

Institutionalization of Environmental Justice Concerns
The first significant federal institutionalization of environmental justice concerns came in 1980 with the passage of the Comprehensive Environmental Response, Compensation and Liability Act (commonly known as Superfund), which provided for the cleanup of toxic-waste sites. In 1986 that law was modified and the Emergency Planning and Community Right-to-Know Act was passed. The Emergency Planning and Community Right-to-Know Act requires government, industry, and citizens to jointly develop emergency plans to deal with releases of toxic chemicals. It also requires industry to report on the quantity and the nature of chemicals being emitted by their plants. The access to information provided by this law has

greatly improved the ability of citizens' groups to monitor industrial activities in their communities.

Concerns about environmental racism led to the creation of an Office of Environmental Justice in the U.S. Environmental Protection Agency in 1992. In 1994 President Bill Clinton signed Executive Order 12898, titled Federal Actions to Address Environmental Justice in Minority Populations and Low-Income Populations, which required all federal agencies to take actions to identify and address "disproportionately high and adverse human health or environmental effects of its programs, policies, and activities on minority populations and low income populations."

Current Environmental Justice Organizations

Regardless of their specific focus, environmental justice organizations define their objective as changing the social order in some manner to solve environmental problems. The means these organizations use include holding government and corporations accountable through democratic processes, such as taking legal action to revoke a corporation's charter and bringing legal suits to end toxic-waste dumping.

Some of these organizations describe their purpose in terms such as "create economic democracy through localized decision-making," "develop grass roots capabilities to involve local citizens in resolution of their communities' environmental problems," or "abolish environmental racism."

The magazines of these organizations focus on strategies and opportunities for change. They report on successes, and on how changes were achieved, and provide guidance designed to enable citizens to take effective action.

To examine the characteristics of organizations informed by this discourse, I examined a sample of nine environmental justice organizations founded between the late 1970s and the early 1990s. The mean of the founding dates was 1982.

Like Deep Ecology organizations, these organizations are very small in comparison to other environmental-movement groups, with an average membership of less than 4000 and an average staff of 9. These groups have even few financial resources than Deep Ecology organizations, with an average annual income of less than $500,000 and total assets of about $150,000. The sources of income of these groups are shown in table 9.2.

Table 9.2
Sources of income of environmental-justice organizations.

	Percentage of income
Public contributions	42.5
Foundation grants	23.4
Government grants	24.3
Program services	8.4
Membership dues	0.7
Other	0.7

Forty-four percent of these groups use education as a primary strategy for influencing policy change; 56 percent use primarily parliamentary means.

Fifty-six percent of the groups have some form of membership participation in their governing; the remaining 44 percent are oligarchic in structure.

Social Learning and Environmental Justice Organizations
Environmental Justice organizations seek redress of the inequitable burden of ecological degradation borne by economically disadvantaged or ethnic minority communities. The environmental justice movement is generally perceived as a decentralized collection of local community groups (Schwab 1994; Bullard 1993).

Networks such as the Citizens' Clearinghouse for Hazardous Waste enable local groups to engage in joint action.[8] Schlosberg (1998: 107–144) argues that these "local groups rarely remain isolated and unconnected." "What makes environmental justice a movement," he comments (p. 121), "are the linkages formed beyond the local." Schlosberg maintains that this network structure creates a stronger movement because it "gives the movement many points of attack, positions from which to argue, and tactics to use, while helping to pool resources efficiently" (p. 135). To the extent that this is actually the case, the environmental justice community embodies many of the characteristics of an open and democratic community, and the increased learning capacity that this entails. However, Schlosberg overstates the uniqueness of networks to environmental justice organizations. His dichotomization between national organizations and networks of environmental justice organizations is a false one. There are many such networks in organizations as varied as Trout Unlimited, the National Wildlife Federation, and the Sierra Club. As the network analysts Emirbayer and Goodwin (1994) and Knoke (1990) have shown, networking among environmental organizations is not a unique characteristic of the environmental justice community. As Tober (1989) has pointed out, networks operate within the field of wildlife management. In addition, the empirical case for Schlosberg's findings is not established. As I noted in chapter 4, a lack of empirical evidence can lead to the "risk that the self-image . . . as grass-roots organizations can be too easily adopted by external observers" (Rucht 1989: 63). Schlosberg's conclusions, which are based on case studies of two environmental justice networks, are suggestive but not definitive.

An examination of the by-laws of the nine Environmental Justice organizations in the sample shows that 56 percent have a participatory structure. While 42.5 percent of the income of these groups comes from the membership, 47.7 percent comes either from foundations or from government grants. Accordingly, these groups have the potential to be influenced by external funding organizations. In addition, those organizations without participatory structures have the potential to be less authentic community representatives. Thus, some Environmental Justice groups are authentic grassroots organizations; others are "Astroturf" organizations.

The discourse of Environmental Justice appears to provide some capacity for authentic movement organizations but does not appear to strongly encourage their formation. Further detailed empirical study using the theories and techniques of network analysis is needed on a larger sample of groups to resolve the question of what the actual practices of the groups based on this discourse are. Thus, the capacity of these groups to facilitate social learning remains an open question. However, it can be said that Environmental Justice is an exclusively anthropocentric discourse. Its concern with nature is limited to examining how ecological degradation affects the human community. This limits the range of issues and concerns addressed by the organizations within this discourse. Environmental Justice cannot provide a basis for the protection of aspects of the natural world that do not affect the well being of the human community. Hence, this discourse cannot inform a cultural practice that could protect biodiversity outside of human-focused utilitarian considerations.

Ecofeminism

The real significance of biological evolution . . . is that humanity is a part of organic nature, subject to laws of development and growth, laws which cannot be broken with impunity. It is our business to study the forces of nature and to conquer our environment by submitting to the inevitable. Only then will we gain control of the conditions which affect our own well-being.
—Ellen Swallow Richards (1912: 39)

"The control of nature" is a phrase conceived in arrogance, born of the Neanderthal age of biology and philosophy, when it was supposed that nature exists for the convenience of man.
—Rachel Carson (1962: 297)

The third alternative discourse, Ecofeminism, ties the development of a patriarchal society and the domination of women by men to the domination of nature by humanity (Merchant 1992: 183–210; Oelschlaeger 1991: 309–316). Nature forms a living web in which humanity is enmeshed. Ecological degradation originates in the treatment of nature as an object to be possessed and dominated rather than as a partner to be cooperated with. Thus, Ecofeminism ties the cultural treatment of nature to the development of a patriarchal society and the domination of women by men. Just as men dominate women, the argument goes, humanity dominates nature. The resolution of ecological problems thus entails a shift from a culture that manipulates and controls both women and nature to a culture of cooperation. The following are key components of this discourse:

• Earth is home for all life and should be revered and nurtured.
• Ecosystem abuse is rooted in androcentric concepts, values, and institutions.
• Relations of complementarity rather than superiority between culture and nature, between humans and nonhumans, and between males and females are desirable.
• The many problems of human relations, and relations between humans and nature, will not be resolved until androcentric institutions, values, and ideology are eradicated.

The Development of Ecofeminism
In 1974 Francoise d'Eaubonne introduced the term *ecofeminism* to describe a new consciousness. She also argued for a shift to ecofeminism as a way to ensure the world's survival by means of a "great upheaval" of male

power, which had brought about agricultural overexploitation and then lethal industrial expansion. To bring about this new world, D'Eaubonne (1974: 178) advocated the egalitarian administration of a world to be reborn in ecofeminism.

Several distinct schools of thought contribute to ecofeminism. *Cultural* ecofeminism sees all recent cultures as fundamentally patriarchal and as denigrating women in significant ways by identifying them with nature. Thus, it is argued, a shift in symbolic structures is needed to regenerate a cultural identification of the Earth Mother and human oneness with nature as desired cultural values. According to *biological* ecofeminism, women develop a psyche centered on empathy, identification, and fusion of the self with nature as a result of their role as reproducer of the family. This psyche is opposed to the male identity, which is based on a distinction between the self and the other, including the self and nature. Nature then becomes an external object to be controlled. *Socio-economic* ecofeminism sees cultural frameworks and the treatment of nature as reflecting different male and female roles in different forms of economic production. In patriarchal capitalism, women and nature are dominated and support the maintenance of male and capitalist activities; consequently, women and nature are devalued and treated as exploitable resources for the use of males and capitalism (Merchant 1995: 142–145).

Women and the Early Environmental Movement

Wolf (1994) traces the initial involvement of women in environmental issues in the United States to the late nineteenth century. Several women's groups, including the Federation of Women's Clubs, the Daughters of the American Revolution, the General Federation of Women's Clubs, and the Woman's National Rivers and Harbors Congress, were active in promoting forest conservation. The California Federation of Women's Clubs actively opposed the damming of Hetch Hetchy Valley (Merchant 1995: 109–136).

Women contributed to the cause of wildlife management and, in one notable action, promoted elimination of the use of feathers in hats. Mazel Osgood Wright, a nature writer, founded the Connecticut Audubon Society (Merchant 1995: 123–128). Rosalie Edge was instrumental in the revitalization of the Audubon Society (Fox 1981: 174–182). She formed the Emergency Conservation Committee in 1929 and worked successfully to transform the Audubon Society from a gun-company-controlled hunting organization to an advocate for the protection of wildlife as beings with their own purposes and ends other than to serve human needs. She sought to "break open the tunnel vision imposed by a man centered view of the universe" (ibid.: 181).

Women also played a major role in the municipal housekeeping movement. They worked with and pressured city officials to take action to through civic improvement campaigns. Homemakers and mothers were seen as having a direct interest in promoting a clean and healthful environment. This cast women as urban reformers in an extension of their domestic role, though it also limited their political action to what was considered appropriate for middle-class married women.

The Consumer-Environment Movement

A significant social movement that can be seen as a precursor to the ecofeminist movement was the Consumer-Environment Movement. This movement grew out of the social feminist movement, which was at its peak between 1890 and 1920. According to Melosi (1981: 117), social feminism was a movement of women who "displayed great sympathy for the disadvantaged, devoted considerable time to civic improvement and hoped to obtain social justice for their sex." One center of this movement was the Hull House settlement in Chicago, founded by Jane Addams.

One of the first persons to apply social feminist ideas to environmental issues was Ellen Swallow Richards. In 1873, she was the first woman to be graduated from the Massachusetts Institute of Technology. After becoming a member of MIT's faculty in 1884, she developed the discipline of Sanitary Chemistry, which was concerned with water, sewage, and air and their effects on human health (Hoy 1980: 179).

Since the home was the basic unit of the physical and social environment of human species, Richards reasoned, any effort to resolve environmental problems must begin there. Her strategy was to create a balanced interaction between humans and nature at the point of contact, where commodities were used, which was the home. Women managed the resources of the home. So the key was to develop information and to provide education to women so that they could manage their homes in harmony with the natural environment (Clarke 1973: 78–79). What was needed, according to

Richards, was the development of a new science which would enable the development of "Right Living Conditions." Right Living Conditions comprised the provision of "pure food and a safe water supply, a clean and disease-free atmosphere in which to live and work, proper shelter, and the adjustment of work, rest, and amusement" (Richards 1912).

Recognizing that women must know some science to defend their homes against the use of harmful products being sold to them by manufacturers, Richards embarked on a career as an activist in developing the science of the home and in advocating this information's widespread application. Her efforts are seen to have created the Consumer-Environment Movement (Clarke 1973: 82–83). This movement had two related components. The first component was known as the Consumer-Nutrition movement. This part of the movement focused on the development of food and drug laws for unadulterated food and medicine to limit the hazards of industrialized food and drug production. To assist in the development of the necessary knowledge of the "Science of Right Living," Richards developed a Woman's laboratory at MIT to train women for science careers (Gottlieb 1993: 216). The second component was to provide women with the knowledge to apply this science. To provide this knowledge, Richards developed an "Environment Education Movement." The aim of this program was to disseminate the "Science of Right Living" so that this information could be used by women in managing their homes. In 1882 she launched her program of educating women in this area with her 1882 book *The Chemistry of Cooking and Cleaning* (Clarke 1973: 96).

In a speech given in November 1892, Richards developed a term for her "Science of Right Living." In her speech, she made the following comment (quoted in Clarke 1973: 120): "For this knowledge of right living, we have sought a new name. As theology is the science of religious life, and biology the science of physical life, so let *Oekology* be hencefor the science of our normal lives, the worthiest of all the applied sciences which teaches the principles on which to found a healthy and happy life." This was the first use of the word Ecology in English. This Oekology became known as the discipline of Human Ecology, which was later on transformed in the term Home Economics.

Following in the example of Ellen Swallow Richards in applying social feminist ideas to environmental issues was Alice Hamilton. Trained as a medical doctor, she took up residence in the 1890s at Hull House. She lived there while working as a professor at the Woman's Medical School of Northwestern University. In this position, she researched the development of industrial diseases, and the role of chemicals in inducing these diseases. Her work gained her national recognition, and in 1919, she became the first woman member of Harvard's faculty. In 1925 she published her landmark book *Industrial Poisons in the United States*. With the publication of this book, she became "the country's most powerful and effective voice for exploring the environmental consequences of industrial activity" (Gottlieb 1993: 51). Hamilton was a tireless advocate for industrial safety and occupational health, including protection for women and minorities who were exposed to industrial chemicals in the course of their work. One of her major research efforts focused on the impact of lead on the human body. Her research led her to be acknowledged as the authority on this subject. Based on her knowledge of the health impacts of lead, she became a leading figure in a campaign to have lead banned from gasoline in the 1920s. Although this campaign was ultimately unsuccessful, it set the stage for further efforts which were taken up again in the 1970s (Neuzil and Kovarik 1996: 129–162).

Women thus worked in many parts of the early environmental movements. In addition, the work of women such as Ellen Swallow Richards and Alice Hamilton set a model for women scholars and activists, thus prefiguring the role played by Rachel Carson in the 1960s and Theo Colborn in the 1990s.

Development of the Ecofeminist Movement

Following the definition of ecofeminism in 1974 by d'Eaubonne, there were a number of similar articles and books that followed. It was in the late 1970s that the development of Ecofeminism as a distinct environmental discourse began in earnest. These ideas were expanded in the works of several authors, among them Sherry Ortner and Rosemary Ruther.[9] With the 1980 publication of Carol Merchant's book *The Death of Nature*[10] and the holding of a major conference on Women and Life on Earth: Ecofeminism in the 1980s (Merchant 1992: 184) that same year, the identity of Ecofeminism as a distinct environmental discourse was established.

This was subsequently followed by the founding of the first ecofeminist organization, World Women in Defense of the Environment, in 1982. Later renamed WorldWIDE, the purpose of this organization is "to promote the inclusion of women and their environmental perceptions in the design and implementation of development and environmental policies; to educate the public and its policymakers about the vital linkages between women, natural resources, and sustainable development; to promote the establishment of a worldwide network of women concerned about environmental management and protection; and to mobilize and support women, individually and in organizations, in environmental and natural resource programs" (by-laws, WorldWIDE Network, October 1992).

In addition to the development of this and other similar organizations (Kelber 1994), women have played an essential role in the development of the grass-roots movement against toxic wastes. In her examination of the activities of minority and white working-class women in this area, Krauss (1993a,b, 1994) shows how their activism "grows out of their concrete, immediate experiences" (Krauss 1993b: 248). Faced with immediate physical threats to the health and well being of their families, these women have acted to organize their communities to meet these threats. However, in the process of challenging the traditional belief systems, these women move beyond these particular problems to challenge the wider issues of "race, class and gender" (ibid.: 259). Hence, women play an important role in Environmental Justice organizations in general, as well as in specifically ecofeminist organizations

Current Ecofeminist Organizations
To examine the characteristics of organizations informed by this discourse, I compiled and examined a sample of five different ecofeminist groups. Most of these organizations were founded around the beginning of the 1980s. The mean founding date is 1982. These organizations are very small in comparison to other environmental-movement groups, with an average membership of 4000 and a staff of two.

Economic information could be obtained on only one of these organizations, so no real analysis of their financial status or income sources could be performed.

As a primary strategy for influencing policy, 60 percent of these groups use education and 40 percent favor parliamentary means. Sixty percent of the groups are oligopolistic; 40 percent have at least limited democracy. In view of the small sample, the data on these characteristics of ecofeminist organizations are preliminary.

Social Learning and Ecofeminism
The ability of the discourse of Ecofeminism to serve as a model for a wide-spread social movement has not yet been developed. There remain open questions that will have to be addressed if this is to occur. First, as in Environmental Justice organizations, most of the political struggles of ecofeminist groups have been over local issues. The development of a larger political vision and agenda will be needed if this discourse is to expand beyond its current use. Second, the prospect of acceptance of an Ecofeminist discourse is highly problematic, given the dominance of patriarchy. A meaningful practice that addresses the transition from a patriarchal to a more egalitarian society will need to be developed to enable a widespread acceptance of this discourse.

Ecotheology

So God created man in his own image, and blessed them, and said to them, "Be fruitful and multiply, and fill the Earth and subdue it; and have dominion over the fish of the sea, and over the birds of the air, and over every living thing that moves upon the Earth."
—Genesis 28

We shall continue to have a worsening ecologic crisis until we reject the Christian axiom that nature has no reason for existence save to serve man.
—Lynn White (1967: 1207)

In 1967, a landmark essay by Lynn White titled "The Historical Roots of Our Ecologic Crisis" appeared in the journal *Science*. In this essay, White argued that the biblical tradition, on which both Judaism and Christianity are based, was the root of the environmental crisis. The Bible created a separation of man from nature. Man was seen as master of, and apart from, the rest of creation. This created a wholly anthropocentric view of nature, in which man was commanded to subdue the Earth. So the exploitation of nature for man's needs was natural and appropriate (Nash 1989a: 194–198).

The remedy for our ecological crisis was clear. If the biblical belief system created a disregard for the natural environment, and led to our ecologic crisis, we needed to develop a new religious viewpoint that would accommodate man to live in harmony with nature. "More science and more technology," White wrote (1967: 1206), "are not going to get us out of the present ecologic crisis until we find a new religion, or re-think our old one." This viewpoint posed a major problem for Western theologians. As Oelschlaeger (1994: 100) noted, "the looming possibility of ecocatastrophe undercuts faith in the notion that God designed the Earth for man." Nonetheless, a number of religious approaches to environmental degradation have arisen since the publication of White's article. Out of these writings, a unique discourse of Ecotheology has emerged. Key components of this discourse include the following:

- Nature is endowed with spiritual value.
- Humanity, as part of nature, has a moral obligation to preserve it intact.
- Religious beliefs that embody this ethic should be developed.
- These beliefs can then inform actions to create an ecologically sustainable society.

The Development of Ecotheology

This application of spiritual ideas to better enable society to deal with the problems of environmental degradation has been realized in several discourses. The spiritual base of Deep Ecology and some aspects of ecofeminism can be seen as attempts to develop alternative spiritual traditions that are not based on either Western biblical tradition or patriarchy.

In addition to these two discourses, the most influential development in the application of religious ideas originated in the environmental justice activities of the black churches. These churches were and continue to remain a major factor in the development of the discourse and practices of the Environmental Justice and Environmental Racism movements. For example, the first protest in 1982 against a toxic landfill in North Carolina was led and organized by a local black church. In addition, the 1987 study on environmental racism performed by the United Church of Christ played a major role in defining the issue of environmental racism.

The activities of these churches have led theologians to recognize the development of a theology of eco-justice. Based in mainline Christian theology with the creation of a just and caring human community, this theology links church perspectives on social-justice issues with environmental concerns and the elimination of environmental racism and injustice. Thus, environmental degradation is seen as a social problem that should be dealt with through political action to realize social and institutional change (Kearns 1996). Thus, the theology of eco-justice can be seen of as part of the larger Environmental Justice discourse.

In the 1980s and the 1990s, a number of new theological viewpoints were developed as a conscious effort to engage Christian beliefs with environmental degradation and to develop an ecological ethic. These viewpoints have been realized in a number of distinct organizations. Although there is a wide variety of religious traditions involved in these organizations, they all share a common orientation toward the revival of religious beliefs as a fundamental means of creating an ecologically sustainable society. Thus, although they have quite different origins, they form a unique discursive community that can be seen as being based in the discourse of Ecotheology (Nash 1989a: 198; Kearns 1996: 57).

The development of ecotheology is based on the premise that Christianity forms a "Great Code" or master narrative of Western culture. Thus, in order to change our social order to incorporate environmental concerns, the Great Code needs to be redefined (Oelschlaeger 1994: 9–11). In addition, pragmatic considerations point to the need to reform religious discourse. Advocates of ecotheology point out that religious discourse is the only discourse that is widely available that can express ethical interests and the public good. In addition, the church going population represents two-thirds of the adult U.S. population (ibid.: 76). Thus, to develop a broad based program of social change, the church-going population needs to become involved in efforts to deal with the process of ecological degradation.

There were a number of early advocates of ecotheology. In 1910, the Conservation Committee of the American Unitarian Association stated: "While the resources of our country have not by any means been exhausted, they have been consumed at an alarming rate. . . . It would seem that the ministers of the churches might help in this matter by bringing it to the attention of their congregations." (AUA 1910: 3–4) Writing in 1915, Liberty Hyde Bailey foreshadowed the modern ecotheologians:

Man has dominion, but he has no commission to devastate: And the Lord God took the man, and put him into the Garden of Eden to dress and to keep it. Verily, so bountiful hath been the Earth and so securely have we drawn from it our substance, that we have taken it all for granted as if it were only a gift, and with little care or conscious thought of the consequences of our use of it. (Bailey 1915: 21)

Another early statement about the religious obligations toward the natural environment occurred in 1939 in a radio address on Jerusalem Radio by Walter Lowdermilk, Assistant Chief of the U.S. Soil Conservation Service. Citing the depletion of forests and soil in China and the Mideast, Lowdermilk maintained that if God had foreseen the destruction of the natural environment caused by man He would have added an eleventh commandment. Then, Lowdermilk made the conservationist creed of Gifford Pinchot into a religious commandment: "Thou shalt inherit the holy Earth as a faithful steward, conserving its resources and productivity from generation to generation." (Lowdermilk, quoted in Nash 1989a: 202)

The theologian Joseph Sittler followed Lowdermilk in the development of ecotheology. In 1954 he published an article titled "A Theology for Earth." This article was followed by an address to the World Council of Churches in 1960. Sittler's argument in both the article and the speech were that man and nature were created as a community, and that there was no separation between them. He advocated following St. Francis in seeing nature as "man's sister" (Nash 1989a: 203).

Interest in the development of an appropriate religious response to environmental problems expanded in the early 1960s among theologians. One result of this concern was the formation of the FAITH-MAN-NATURE Group in 1964. Sponsored by the World Council of Churches, this group had as its purpose to "stimulate and assist in the orientation and encouragement of a better developed theology of Nature and of man in relation to Nature" (FMN Papers #1 1967: 11). This group held a series of meetings in the mid 1960s and the early 1970s to explore the relationships between humanity and the natural environment. The group disbanded in 1974 as a result of financial problems.

Theological concern over the role of religion and environmental degradation continued to expand in the 1970s and the early 1980s (Nash 1989a: 211–214). In 1985 the North American Conference on Christianity and Ecology was founded. The purpose of this organization was to assist theologians in developing a common vision which would "elucidate Christianity's ecological dimension" (Kearns 1996: 59). At the first, and subsequent meetings, theological disputes between those seeing a need for a traditional interpretation of the Bible and those seeking the transformation of these perspectives erupted. As a result, in 1988, a group of theologians split off from the NACCE to form the North American Coalition on Religion and Ecology. Made up of liberal Protestant and Catholic theologians, its purpose is to facilitate a dialogue between ecumenicists, ecologists, and economists on resolving ecological problems. This split grew into two different theological perspectives: Christian Stewardship and Creation Spirituality (Kearns 1996).

Christian Stewardship

The perspective known as Christian Stewardship focuses on an evangelical interpretation based on a biblical mandate to care for God's creation. Based in conservative Christian theology, it seeks to create a Christian ecological ethics. This ethics would be created through development of a new interpretation of biblical strictures, thus creating a moral commandment to preserve God's creation. It is basically a reformist perspective, and makes minor adjustments in Christian theology to accommodate environmental concerns. It still sees God as a transcendent being, and human nature as fallen, sinful, and in need of redemption.

The first comprehensive statement of this perspective was made by the theologian Francis Schaeffer in his 1970 book *Pollution and the Death of Man: The Christian View of Ecology*. Schaeffer (ibid.: 76) argues that, on the basis of the Bible, man has an obligation to care for creation: "Christians, who should understand the creation principle, have a reason for respecting nature, and when they do, it results in benefits to man. Let us be clear: it is not just a pragmatic attitude; there is a basis for it. We treat it with respect because God made it."

This was followed by Cal DeWitt's 1991 book *The Environment and the Christian*. DeWitt expands Schaeffer's argument by integrating scientific knowledge about specific ecological problems and biblical strictures against the practices that create ecological problems (Oelschlaeger 1994: 131). DeWitt also argues that the biblical scriptures are not the root cause of ecological degradation. Rather it is human failure to follow the practices that are prescribed by scripture (Kearns 1996: 59).

Creation Spirituality

The second theological perspective in ecotheology is defined by the term *creation spirituality*. In this perspective, the Christian biblical tradition is seen as beyond redemption in light of ecological problems. Instead, there is a need to go beyond the Christian tradition to develop alternative notions of the creation. This approach seeks to create a new synthesis of religion and science. This new synthesis is seen to enable the transformation of society to act within ecological limits. Creation spirituality focuses this reorientation of religious thought around the idea of a pantheistic creation, and the interconnections between human life and the global ecosystem.

One of the important theologians of creation spirituality is Matthew Fox. His 1983 book *Original Blessing* established his agenda to modify and develop a theological alternative to the Christian tradition. One core of his work is to "overcome the dualisms of the Western worldview so that we can see the creation as a whole" (Kearns 1996: 61). Fox accepts that all religions are revelations of the sacred in different cultural contexts, a position that has resulted in his removal from a teaching position at Catholic University. He also maintains that the crucifixion of Christ can serve as a metaphor of how the Earth is treated by humanity (Oelschlaeger 1994: 167–169).

Another major theologian in this perspective is Thomas Berry. For him, the merging of Genesis with scientific knowledge is a new revelation, or a "new story," of humanity and all life on Earth. This "new story" removes humanity from a position of privilege in the universe. Instead, using the knowledge of ecology, humanity is seen as only a part of a whole, and that we need to use our reflective powers to create an ecologically sustainable society in accordance with this new revelation (Oelschlaeger 1994: 165–167).

Development of Collective Action

One of the first institutions based on this discourse is the Au Sable Institute, which was reorganized in 1979, under the leadership of Cal DeWitt, as an institute for environmental stewardship (Frame 1996: 82). Its purpose "is the integration of knowledge of the creation with biblical principles for the purpose of bringing the Christian community and the general public to a better understanding of the Creator and the stewardship of God's creation."

In the mid 1980s, a new movement emerged in religious communities that had traditionally not been involved in environmental issues. In 1988, a conference on Peace, Justice and the Integrity of the Creation brought together many evangelical groups that shared an interest in the natural environment. This was followed in the late 1980s and the early 1990s by a series of proclamations by a number of major religious groups on the religious obligation to protect the natural environment. In 1989, Pope John Paul II wrote an encyclical titled The Ecological Crisis; A Common Responsibility. That same year, the American Baptist Churches approved a statement on religion and the environment titled Creation and the Covenant of Caring. In 1990, the Evangelical Lutheran Church in America issued its statement in this area: Basis for Our Caring. These statements culminated in July of 1991 in the development of a unified statement by leaders of 24 major U.S. religious groups, titled Statement by Religious Leaders at the Summit on the Environment.[11]

By 1993 virtually all the major churches in the United States had issued proclamations on environmental degradation. The only major denominations not to have made such proclamations were the National Baptist Convention and the Church of God in Christ. The Church of Jesus Christ of Latter-Day Saints is the only major church that has formally stated opposition to including ecological concerns into its ministry (Oelschlaeger 1994: 204).

The National Religious Partnership, formed in 1993, unites the United National Council of Churches, the U.S. Catholic Conference, Consultation on the Environment and Jewish Life, and the Evangelical Environmental Network in an organization focused on developing and implementing religious approaches to combating environmental degradation.

These actions have led some social theorists to argue that we are now experiencing a religious revival within Christianity (Kearns 1996; Oelschlaeger

1994). McLoughlin (1978: 10) describes religious revivals as "periods when the cultural system has had to be revitalized in order to overcome jarring disjunctions between norms and experience, old beliefs and new realities, dying patterns and merging patterns of behavior."

Current Ecotheology Organizations

The organizations based on this discourse are just emerging, and only a few small organizations exist. Hence, generalizations about their income, strategies, or organizational forms would be premature. In addition, several of the groups are subsidiary efforts of established religious groups. In these cases, determining their structure and funding as a distinct entity is virtually impossible. As these groups emerge and develop a more established identity, further analysis may become possible.

Social Learning and Ecotheology

The discourse of Ecotheology is also still emerging. However, some initial observations can be made:

• The process of change to an ecologically sustainable society is based upon individual conversion to this viewpoint. Subsequently, a comprehensive social and political theory of transition has not been addressed by this discourse.

• One is given pause to think about what type of practices this type of religious discourse would enable. Ecclesiastical organizations are generally not democratic in form.

• Ecotheology tends toward being an anthropocentric discourse. It calls on humans to value nature, not for its own sake, but rather because of its origins in God's creation. This ties reverence toward nature to belief in God. This limits its applicability and acceptance outside of the community of believers.

10

The Dynamics of the Environmental Movement

Though the decentralized network structures of new social movements have been often emphasized, there is no clear evidence that such structures are always present.
—Dieter Rucht (1989: 63)

One of the major hopes of the environmental movement is that ecological degradation might be reversed through a rapid increase in society's capacity for social learning. This capacity is reliant on a strong and democratic public sphere. Social movements are key actors in the creation and maintenance of this type of public sphere. This role is premised on social-movement organizations' being located in a structural position not controlled by the logic of capital accumulation or that of the administrative state but rather formed and governed by citizens through free and open discussion. By mobilizing citizens and by providing legitimate and effective representation of their needs, social-movement organizations can act as catalysts for effective political demand for change. However, if these organizations are not authentic representatives of communities, or if they restrict the flow of communication from citizens to the public sphere, their independence is compromised and their ability to foster social learning is limited. Thus, there is a need to identify conditions that foster or inhibit an organization's ability to communicate lifeworld concerns into the public sphere.

The literature on social movements has identified a number of factors that can limit the ability of a movement to foster social learning by systematically distorting communication or by limiting the participation of members. Certain discourses can mask consideration of certain phenomena, thus limiting the range of options considered. In addition, the discursive frame influences the structure of organizations. Thus, some discourses may foster the development of oligarchic institutions. Finally, the means of recruiting members and raising funds have an important effect on individual commitment to movement organizations. Thus, different means of fundraising can either limit or increase the participation of members in an organization. Finally, the sources of funds can have an important impact on the independence and management of a movement organization.

The previous chapters portrayed a number of U.S. environmental organizations in terms of their historical development, the discursive frames that define them, the resources mobilized by them, their organizational structures, and the strategies they use to bring about change. In this chapter I combine these individual portraits to develop a comprehensive picture of the entire environmental movement, seeking to identify the social factors that can assist in developing a democratic and sustainable society by fostering social learning.

Historical Evolution of Organizations

Different environmental movements have been active in the United States in different periods. Although there were movements for urban parks, local game protection, and urban sewer and water systems in the middle of the nineteenth century, the most active periods of ecological politics in the United States were 1890–1915 and 1962–1975. In the 1980s, a number of new movements and organizations arose from alternative discourses. With the exception of soil-conservation concerns and the development of the discipline of wildlife management, there was very little activity in ecological politics from the 1920s to the 1960s.

As is evident in table 10.1, organizations based in the Preservation discourse are the oldest, with a mean founding date of 1933. Organizations founded in the discursive frames of Wildlife Management and Conservation are the next oldest. Reform Environmentalist organizations' founding dates are clustered around 1970. Organizations based on the alternative discourses of Environmental Justice, Deep Ecology, and Ecofeminism were founded in the early 1980s.

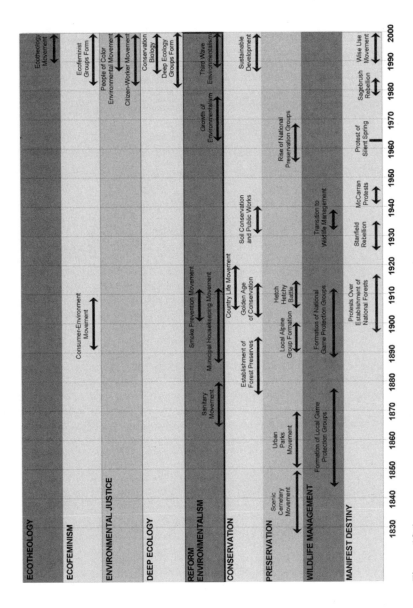

Figure 10.1
A time line of U.S. environmentalism.

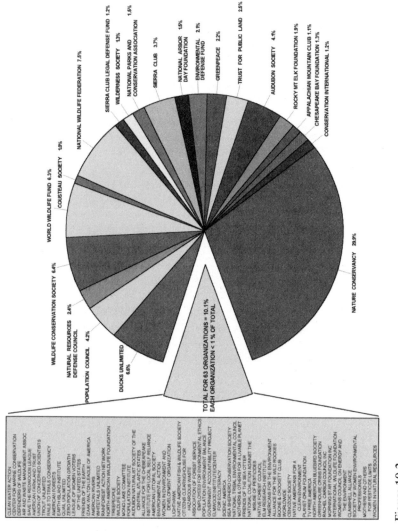

Figure 10.2
Income distribution of U.S. environmental organizations.

Table 10.1
Founding years of 87 environmentalist organizations by discourse.

	Mean	Standard deviation	Median
Wildlife Management	1947	36	1948
Conservation	1957	39	1972
Preservation	1933	35	1935
Reform Environmentalism	1969	18	1971
Deep Ecology	1982	8	1985
Environmental Justice	1985	6	1984
Ecofeminism	1982	4	1982
All	1962	30	1973

Characteristics of Sectors of the Movement

In this section I use data on 87 organizations[1] to estimate the overall dimensions of the various sectors of the environmental movement and to evaluate the various environmentalist discourses on the basis of number of organizations, number of members, size of staff, annual income, and net worth.

It is important to note here that the Nature Conservancy has by far the largest economic resources and the largest staff of any U.S. environmental organization. This preservationist organization plays a unique role in the movement. Because its major activity is acquiring and managing ecologically sensitive land areas, its economic resources (both annual income and net worth) are extraordinary. It accounts for nearly 30 percent of all the income of the 84 organizations for which income data were available, its staff of 2,000 constitutes 32 percent of the total for the sample, and its net worth amounts to 66 percent of the total for the sample. The size of these resources tilts any analysis of patterns in the movement. Accordingly, each of the analyses below includes a discussion of the pattern that results with the Nature Conservancy included and with it excluded.

Organizations by Discourse

As table 10.2 shows, the largest number of organizations (31) have Reform Environmentalism as a primary discourse. Organizations with Preservation as a primary discourse are next, numbering 18. The other discourses are represented by roughly equal numbers of organizations.

Table 10.2
Distribution of organizations by discourse.

	Number of organizations	Percentage of total
Wildlife Management	7	8.1
Conservation	8	9.2
Preservation	18	20.7
Reform Environmentalism	31	35.6
Deep Ecology	8	9.2
Environmental Justice	9	10.3
Ecofeminism	6	6.9

A great deal of effort was expended to identify and to obtain data on organizations that were not within the Reform Environmentalist frame; however, additional data on the Environmental Justice, Deep Ecology, and Ecofeminism organizations are still needed.

Number of Members by Discourse
There is considerable variation among organizations in the methods they use to count their members. Since number of members is a measure of an organization's political strength, there is an incentive to exaggerate it. In any case, an organization's reported membership may be indicative of its ability to mobilize resources.

The total membership of environmental organizations in the United States is estimated to be between 19 million and 41 million. The 87 organizations in the sample represent between 29 percent and 63 percent of that total.

Table 10.3 presents the data on the 87 organizations in the sample.

Staff Size by Discourse
In the data on staff size (table 10.4), the sheer size of the Nature Conservancy is evident in the row for preservationist organizations. Without the Nature Conservancy included, the total number of staff employees in preservationist organizations drops to 1061, the average number per organization drops from 235 to 88, and the Wildlife Management organizations have the largest staffs.

Table 10.3
Distribution of members by discourse.

	Members	Percentage of total membership	Average per organization
Wildlife Management	5,168,120	42.3	861,353
Conservation	1,063,000	8.7	132,575
Preservation	3,977,115	32.6	248,570
Reform Environmentalism	1,915,400	15.7	95,770
Deep Ecology	62,400	0.5	12,480
Environmental Justice	18,775	0.2	3,755
Ecofeminism	7,830	<0.1	3,915
Total	12,212,640		182,650

Table 10.4
Distribution of staff by discourse.

	Size of staff	Percentage of total staff	Average per organization
Wildlife Management	859	13.7	172
Conservation	306	4.9	44
Preservation	3,061	48.7	235
Reform Environmentalism	1,971	31.4	86
Deep Ecology	38	0.6	10
Environmental Justice	46	0.7	9
Ecofeminism	4	<0.1	2
Total	6,285		95

Annual Income by Discourse

As can be seen in table 10.5, the 87 organizations in the sample account for nearly 40 percent of the annual income of the U.S. environmental movement (approximately $2.7 billion). With the Nature Conservancy's $322 million omitted, the total income of the Preservationist organizations drops to $297 million and the average income of those organizations drops from $34.4 million to $17.5 million.

Even without the Nature Conservancy, the Preservationist organizations have the largest annual income; the Wildlife Management organizations

Table 10.5
Distribution of annual income by discourse.

	Total income (millions)	Percentage of total income	Average income per organization (millions)
Wildlife Management	$181.77	16.9	$25.97
Conservation	$59.92	5.6	$7.49
Preservation	$619.10	57.5	$34.39
Reform Environmentalism	$206.72	19.2	$6.89
Deep Ecology	$3.65	0.3	$0.52
Environmental Justice	$3.78	0.4	$0.47
Ecofeminism	$1.59	0.1	$0.27
Total	$1076.53	$12.00	

are a close second. The Conservationist and the Reform Environmentalist organizations are roughly equal in income. The combined economic resources of Environmental Justice, Deep Ecology, and Ecofeminism organizations amount to just over 1 percent of the total income.

Net Worth by Discourse
As table 10.6 shows, Preservationist organizations have nearly all of the net assets of the organizations sampled. With the Nature Conservancy's net worth of $1.034 billion subtracted, the total for the Preservationist organizations drops to $301.6 million and the average per organization drops to $17.73 million. This still amounts to 56 percent of the total net worth of the organizations in the sample. The net worth of the Environmental Justice, Deep Ecology, and Ecofeminism organizations amounts to 0.6 percent of the total.

Primary Strategies
As Lofland (1989) showed,[2] different implicit theories of how to bring about social change underlie the activities of organizations within the various discourses. In this book, I used the following categories.

transcendence   Bring about a rapid shift in consciousness through charismatic and aesthetic demonstrations.

Table 10.6
Distribution of total net worth by discourse.

	Total net worth (millions)	Percentage of total net worth	Average net worth per organization (millions)
Wildlife Management	$40.2	2.6	$7.5
Conservation	$60.0	3.8	$5.7
Preservation	$1335.6	85.2	$74.2
Reform Environmentalism	$130.1	8.3	$4.6
Deep Ecology	$0.8	<0.1	$0.1
Environmental Justice	$1.0	0.4	$0.1
Ecofeminism	$1.0	0.1	$0.3
Total	$1569.66	$18.9	

education   Instruct the population in necessary changes through development and dissemination of publications and materials.

parliamentary   Work through existing market, legislative, and bureaucratic processes to create incremental changes.

protest   Force change by noncooperation and/or by disruption of regular routines.

prophecy   Bring about deep moral regeneration of society by creating separate communities that rebuild self and society.

As table 10.7 shows, the overwhelmingly dominant strategies for change among the organizations sampled are education and parliamentary. Only 8 out of 83 organizations (9.6 percent) rely primarily on tactics that can be characterized as being outside normal political channels.

As table 10.8 shows, organizations based in the discourse of Wildlife Management or that of Ecofeminism follow almost exclusively an education strategy. Conservationist, Preservationist, Reform Environmentalist, and Environmental Justice organizations use a mix of education strategies and working through established social institutions. Deep ecology organizations use a wide range of strategies, including a high degree of protest.

Number of Organizations by Organizational Form

For purposes of this study, "organizational form" pertains to the structure by which decisions are made. Organizations with a democratic form are

Table 10.7
Distribution of primary organizational strategies (Goodman and Kruskal τ strategy dependent on discourse value = 0.107, significance = 0.044).

	Number	Percentage
Transcendence	1	1.2
Education	41	49.4
Parliamentary	34	41.0
Protest	6	7.2
Prophecy	1	1.2
Total	83	

Table 10.8
Distribution of discourses and strategies.

	Transcendence	Education	Parliamentary institutional reform	Protest	Prophecy
Wildlife Management	0	5 (83.3%)	1 (16.7%)	0	0
Conservation	0	4 (57.1%)	3 (42.9%)	0	0
Preservation	0	9 (52.9%)	8 (47.1%)	0	0
Reform Environmentalism	0	15 (48.4%)	14 (45.2%)	2 (6.5%)	0
Deep Ecology	1 (12.5%)	1 (12.5%)	1 (12.5%)	4 (50.0%)	1 (12.5%)
Environmental Justice	0	4 (44.4%)	5 (55.6%)	0	0
Ecofeminism	0	3 (60.0%)	2 (40.0%)	0	0

better able to transmit the lifeworld impulses from their members than organizations with more restrictive organizational forms. As I noted in chapter 4, this analysis identified four organizational types:

oligarchy   The organization is governed by a board of directors, a self-replicating mechanism that elects the officers of the organization. No provisions for input by individual members exist in the bylaws.

representative   Members of the organization can elect representatives of their local chapters. These representatives then participate in the selection of the organization's board of directors, officers, and policies.

limited democracy   The organization is governed in part by a board of directors and in part by the members. Individual members can nominate and/or elect some of the members of the board or some of the officers of the organization. However, certain aspects of organizational control are specifically delegated to the board of directors.

democracy   The organization is governed by its members. The board of directors and the officers are nominated and elected by the membership. Policies of the organization can be debated and voted upon by individual members.

This scheme is limited. It identifies only the formal structure of an organization as indicated in its by-laws. If its by-laws specify a democratic structure, an organization is considered to have one even if that is not reflected in its actual practices. Since formally democratic but practically oligarchic organizations are recorded as democratic, the estimated number of democratic organizations should be seen as the upper limit on the extent of democracy in the movement.

As table 10.9 shows, more than 61 percent of the organizations in the sample are oligarchic. The next most frequent type of organizational form is limited democracy, with 20 percent. Only 11.2 percent of the organizations sampled are democratic.

Membership Size by Organizational Form
Table 10.10 shows the distribution of membership size by organizational form. Here the representative organizations have the largest membership. However, oligarchies account for 38 percent of the membership. Democracies and limited democracies together account for about 19.3 percent.

Table 10.9
Distribution of organizations by organizational form.

	Number of organizations	Percentage of total
Oligarchy	49	61.3
Representative	6	7.5
Limited democracy	16	20.0
Democracy	9	11.2
Total	80	

Table 10.10
Distribution of membership size by organizational form.

	Total number of members	Percentage of total	Average per organization
Oligarchy	4,624,230	38.04	132,121
Representative	5,188,075	42.67	864,679
Limited democracy	1,727,120	14.21	123,366
Democracy	618,215	5.08	123,643
Total	12,157,640		151,971

Staff Size by Organizational Form

Staff size follows a much different pattern. The difficulty in using staff and income figures to measure the extent of democracy in environmental organizations is, as was mentioned above, the overwhelming influence of the Nature Conservancy, which is categorized as a limited democracy.[3]

The distribution of staff resources by organizational form is shown in table 10.11. With the Nature Conservancy included, the limited democracies have the most staffers. Without the Nature Conservancy, the oligarchies have the most.

Total Annual Income by Organizational Form

With the Nature Conservancy included, limited democracies have the most annual income (table 10.12). Without the Nature Conservancy, oligopolies have 53 percent of the income; democracies and limited democracies together have 22.2 percent.

Table 10.11
Distribution of staff by organizational form.

	Total number	Percentage of total	Average per organization
Oligarchy	2,147	35.8	65
Representative	810	13.5	135
Limited democracy	2,676	44.6	206
Democracy	362	6.0	91
Total	5,995		76

Table 10.12
Distribution of annual income by organizational form.

	Income (millions)	Percentage of all income	Average income per organization (millions)
Oligarchy	$348	35.7	$7.4
Representative	161	16.5	26.9
Limited democracy	414	42.4	25.9
Democracy	53	5.4	6.7
Total	$976		12.4

Total Net Worth by Organizational Form

Table 10.13 shows the data on total net worth by organizational form. The influence of the Nature Conservancy is again evident here. Without that organization, the net worth of the limited democracies drops to 21.65 percent of the total and that of the oligarchies increases to 62.35 percent.

Organizational Form by Discourse

The data on organization form by discourse are tabulated in table 10.14. As can be seen, organizations based in the discourses of Preservation, Reform Environmentalism, and Ecofeminism show a strong trend toward fostering the creation of oligarchic institutions, and all the organizations based in Deep Ecology are oligarchic. Only Conservation, Wildlife Management, and Environmental Justice organizations show a trend toward more open and representative organizations.

Table 10.13
Distribution of net worth by organizational form.

	Net worth of all organizations (millions)	Percentage of total net worth	Average net worth per organization (millions)
Oligarchy	$247	17.3	$5.6
Representative	35	2.5	5.8
Limited democracy	1,120	78.3	70.2
Democracy	29	2.0	4.8
Total	$1,431	$18.1	

Table 10.14
Distribution of organizational forms among discourses (Goodman and Kruskal τ organizational form dependent on discourse value = 0.106, significance = 0.045).

	Oligarchy	Representative	Limited democracy	Democracy
Wildlife Management	2 (33.3%)	3 (50.0%)	1 (16.7%)	0
Conservation	3 (42.9%)	1 (14.3%)	3 (42.9%)	0
Preservation	10 (62.5%)	0	4 (25.0%)	2 (12.5%)
Reform Environmentalism	21 (67.7%)	1 (3.2%)	6 (19.4%)	3 (9.7%)
Deep Ecology	6 (100.0%)	0	0	0
Environmental Justice	4 (44.4%)	1 (11.1%)	1 (11.1%)	3 (33.3%)
Ecofeminism	3 (60.0%)	0	1 (20.0%)	1 (20.0%)

Summary of Data

The data show that there are significant differences among the sectors of the environmental movement. Organizations based in the traditional discourses of Wildlife Management, Conservation, Preservation, and Reform Environmentalism have substantial memberships and economic resources and large staffs. Organizations based in the alternative discourses of Deep Ecology, Environmental Justice, and Ecofeminism have few members, small staffs, and nearly no economic resources.

The vast majority of organizations follow the traditional strategies: education and parliamentarian action. Only a few organizations, based in alternative discourses, follow nontraditional or disruptive strategies.

A majority of the organizations in the overall sample are oligarchic, as are a majority of the organizations based in Deep Ecology, Preservation, Reform Environmentalism, and Ecofeminism.

Funding

Environmental organizations' sources of funding influence their choices of strategy and their organizational characteristics. Though accurate data on the distribution of environmental organizations' income sources are not available, a basis for good estimates exists. On the basis of a survey of 265 leaders of environmental organizations, Snow (1992: 63) created the estimated distribution of income for environmental organizations shown here in table 10.15. In this analysis, the largest source of income was membership dues and individual members' contributions. The second largest source was grants from foundations. To further check the reliability of this estimate, I examined data from the aforementioned sample of 87 organizations (table 10.16) and found members' contributions to be the largest source of income. In these two analyses, public support and membership dues account for between 43 percent and 48 percent of the organizations' income. Grants from outside organizations, including the federal government and foundations, were the second largest source. The combined totals for grants from all external funding sources ranged from 22 percent to 29 percent of total income.

The importance of the various sources of funding in maintaining environmental organizations is reflected in Snow's survey (ibid.: 64). Snow

Table 10.15
Estimated sources of environmental groups' income (source: Snow 1992: 63).

Membership dues	24%
Individual contributions	19%
Sales	7%
Federal grants and contracts	4%
Foundation grants	21%
State grants and contracts	2%
Other contracts	2%
Corporate gifts	4%
Capital assets	3%
User fees	3%
Other	11%

Table 10.16
Funding sources of U.S. environmental organizations, 1995.

	Total contribution (millions)	Percentage of total
Public support	$428.27	39.8
Program services	$91.25	8.6
Membership dues	$83.99	7.8
Government grants	$92.69	8.6
Foundations grants	$146.77	13.6
Other	$233.14	21.7
Total	$1076.11	

Table 10.17
"Crucial" funding sources as rated by leaders of environmental organizations.

	Percentage of leaders rating source crucial
Small contributions from individuals	42
Large grants from foundations	39
Small grants from foundations	28
Large contributions from individuals	23
Gifts from corporations	10

asked the 265 leaders in his sample to rate the importance of each of five funding sources on a scale ranging from 1 ("unimportant') to 4 ("crucial"). The percentage rating each source of funding "crucial" is shown in table 10.17.

Mass Mailings

In the early 1980s there was a shift from grassroots-based fund raising to fund raising by means of mass mailings (Hayes 1986). Reliance on mass mailings affects an environmental organization in three ways.

First, it creates a group with shallow and tentative membership support. Social-movement organizations that use this method of fund raising have little or no member involvement. There is virtually no face-to-face contact with other members in a common setting. The membership consists mostly of "checkbook" members, affiliated with the group only by virtue of monetary contributions. The membership developed through mass mailing has only very tenuous links to a particular organization. As Godwin and Mitchell note (1984: 837), mass mailings fail to generate long term community support: ". . . the key implication of the above findings concerns the difficulties that interests supported largely by direct mail associations will have in organizing for local political struggles and in keeping members active for lengthy political battles." This results in episodic swings in membership and in funding (ibid.: 836). One important example of this process involves Greenpeace. This organization is highly reliant on mail donations. In the early 1990s, it experienced a dramatic shift in its mailing contributions. The result was a reduction in its budget by one third from 1990 to 1991 (Martel and Holman 1994: 68). Other environmental organizations have also experienced these types of membership and funding shifts (Carney and Moore 1992: 30). Another reflection of the importance of direct mail for environmental organizations in reflected in the selection of Roger Schlickeisen as president of Defenders of Wildlife. One of his major qualifications for this job was his background as a direct mail specialist (Gifford 1991).

Second, it shifts the issue focus of the organization. Because of the importance of direct mail income, the managers of the direct mail operations are frequently at meetings where issues and strategies are discussed. The question of whether or not the issue campaign will "work" in direct mail campaigns often has a decisive influence over what issues are pursued (Dowie

1995: 44). For a direct mail appeal to attract sufficient contributors, it must appeal to moderate reformers. Thus, it is unlikely to pursue more radical causes or strategies for fear that this might adversely affect mail contributions (Dowie 1995: 43). Thus, the results of direct mail become, in the words of direct mail specialist, Jeffrey Gillenkir, "tacit opinion polls that organizations rely on to guide polity and programs" (quoted, Dowie 1995: 42–43).

Third, the ability of the membership to participate in the organization is decreased. For information regarding the organization's actions other than what can be found in the organization's own self-serving publications, the members are entirely dependent on a remote and impersonal national leadership; thus, they are vulnerable to manipulation. Reliance on mass mailings also tends to concentrate power in the central staff of the organization, and to make the "members" into mere contributors without any meaningful participation (Hayes 1986: 135–136). This decreases the organization's capacity for political mobilization and its ability to change attitudes and beliefs of individuals through active, face-to-face participation in a social movement.

External Organizations

The second largest source of income for environmental organizations is contributions from external organizations. As the earlier analysis showed, approximately 25 percent of the income of environmental organizations comes from these sources. Most of the contributions come from philanthropic foundations. Government and corporate grants make up only a small percentage of the income from external organizations.

Foundations have a long history of contributing to environmental organizations. One of the earliest foundations to make systematic contributions for ecological issues was the Andrew W. Mellon Foundation. Its predecessor, the Old Dominion Foundation, established and funded the Conservation Foundation in 1948. The Ford Foundation joined in this effort in 1965 with its Resources and Environment Program. In 1969 the trustees of the Rockefeller Foundation added the environment to the foundation's interests (Wing 1973: 47–51). Since that time, the John D. and Catherine T. MacArthur Foundation and the Pew Charitable Trusts have added environmental issues to their agendas for major funding.

In 1987, a number of foundations banded together to form the Environmental Grantmakers Association. This organization has grown to a membership of 183 foundations (Cockburn and St. Clair 1994: 761). Its purposes are as follows (EGA 1999):

To promote recognition that the environment and its inhabitants are endangered by unsustainable human activities.

To facilitate communication, foster cooperation, and develop collaboration among active and potential members.

To increase awareness of the relationships between environmental grantmaking and other areas of grantmaking and to encourage all types of philanthropic programs to support environmentally-related activities.

To provide the means by which members can improve their effectiveness as grantmakers.

To increase the resources available to address environmental concerns.

To communicate grantmaker interests and activities to grantseekers and other interested parties.

To help members to set examples of environmental responsibility through their policies and operations, and to encourage others to do likewise.

To realize these purposes, the Environmental Grantmakers Association schedules annual retreats at which representatives of the various foundations are briefed on selected environmental issues by leading government officials, members of the U.S. House of Representatives, and representatives of environmental organizations. The EGA also works to coordinate funding through the formation of working groups focusing on specific issue areas (such as the Sustainable Agriculture Working Group, the Grantmakers Network on the Economy and the Environment, the Funders Collaborating on Minorities and the Environment, the Great Lakes Working Group) and through the formation of interest groups with names such as Resource Recovery/Enterprise Development, Toxics/Health, and Land Use/Growth Management.

Foundation income plays a major role in the startup and the maintenance of environmental organizations. Walker (1991) has documented the importance of outside funding for the startup of citizens' groups. There is significant evidence that the trend identified by Walker is present in the development of environmental organizations. Foundation support played an important role in the founding of several prominent environmental organizations. For example, the Ford Foundation was directly involved in the founding of the Natural Resources Defense Council, the Environmental Defense Fund, and the Sierra Club Legal Defense Fund (Robinson 1993: 39).

Jenkins (1987, 1989) conducted a more detailed analysis of the role of foundation support for social-movement organizations, considering data from 131 foundations for the period 1953–1980 in relation to the development of social protests and movement organizations during that period and finding that "movement and funding take-off did coincide" (1989: 305).

In addition to its role in the startup of environmental organizations, outside funding plays an important role in maintaining such organizations (Cockburn and St. Clair 1994: 761; Walker 1991: 82). Aside from the fact that about 25 percent of the income of environmental organizations comes from external organizations, 39 percent of environmental-organization leaders polled by Snow (1992: 64) rated "large foundation grants" crucial for the operation of their organization.

Impacts of Foundation Support

Although foundations provide only about 25 percent of the funding of environmental organizations, their influence on these organizations is considerable (Adler 1995: 87). One of the main reasons for this is that foundation money is given for project specific functions, and the actions of the organization in fulfilling the grant requirements are monitored by the foundation. Membership support, though larger, is diffuse; it is not targeted on specific projects. In addition, the desire to ensure a continuing flow of funds from a foundation inclines an environmental organization to follow the expressed desires of a foundation representative.

The influence of external funding creates a dynamic that can be seen as financial steering of the environmental organization. Foundation funding draws social-movement organizations into a network of power and control, thus limiting the range of organizational forms and goals. Foundation funding also allows a foundation to gain direct influence over a movement organization through participation on the board of directors (Colwell 1993: 105). Thus, the funding of movement organizations by foundations shifts the locus of control of the organization from the membership to the financial patrons (Hayes 1986; Schmitter 1983; Wilson 1990; Jenkins 1989).

"By deciding which organizations get money," Pell observes (1990: 25), "the grant-makers help set the agenda of the environmental movement and influence the programs that activists carry out." This reduces an organization's independence and affects its issue focus, its organizational form, the practices it uses to effect social change, its efforts at grassroots mobilization, and its level of political legitimacy.

Shift in Issue Focus

Foundations are capable of shifting the activities of a social-movement organization by making their financial aid conditional. Specific conditions can require a movement organization to take up certain issues or to engage in certain practices that it normally would not have taken up or engaged in. "By deciding which organizations get money," Pell notes (1990: 255), "the grant-makers help set the agenda of the environmental movement and influence the programs that activists carry out."

Another way in which the issue focus of social-movement organizations is shifted is through the solicitation of grant proposals from environmental organizations. The foundations decide what issues will be addressed, and how. Then they make their financial aid contingent on which environmental organizations can best accomplish this task based on their proposal. "Foundations," Dowie notes (1995: 50), "have been meeting to decide where the environmental movement should be going. They create multi-million-dollar mega-projects and invite organizations to apply for grants to activate them."

To further examine foundation support of environmental organizations, I examined specific data on foundation grants for the 87 organizations in the sample. The results of this analysis are shown in table 10.18. As with the previous analysis, the grant distribution is skewed by the amount of funding the Nature Conservancy receives. In 1994 this organization received approximately $97 million in foundation grants—66 percent of the total grants awarded in the sample. This overwhelming foundation support for one organization distorts any analysis of these data. Without the Nature Conservancy, foundation funding of Preservationist organizations drops to $15.14 million (30 percent of the total). The average grant to a Preservationist organization drops to $890,000. Reform Environmentalism becomes the largest foundation-funded discourse, with nearly 44 percent of

Table 10.18
Distribution of grant income by discourse.

	Grant income (millions)	Percentage of all grant income	Average grant income per organization
Wildlife Management	$2.2	1.53	0.320
Conservation	$9.8	6.70	1.229
Preservation	$12.1	76.41	6.230
Reform Environmentalism	$21.8	14.87	0.704
Deep Ecology	$0.7	0.44	0.081
Environmental Justice	$0.2	0.06	0.096
Ecofeminism	$0.0	0.00	0.0
Total	$146.7		1.696

the foundation funding. Reform Environmentalism's top-down, scientific analysis of ecological problems obscures their political and economic causes and thus does not threaten existing power relationships.

In essence, the pattern of foundation funding gives certain movement organizations and discourses a larger voice in the policy arena than other movements and discourses. The externally funded discursive frames become more highly represented in the public space than their actual support by citizens would allow.

Shifts in Organizational Form

Foundation funding brings with it responsibilities that require a movement organization to adopt a centralized accounting system and executive management of program activities.[4] In addition, the development of the necessary technical staff capabilities expands staff management of organizational activities. As a result, members' participation declines and the staff's control over the organization increases (Powell and Friedkin 1987: 191). Thus, foundation funding brings a shift from more open and democratic forms of organizational control toward oligarchy.

Table 10.19 illustrates the relationship between foundation support level and internal organizational form. Here the influence of the funding of the Nature Conservancy shows in the high levels of foundation funding of

Table 10.19
Distribution of grant income by organizational form.

	Grant income (millions)	Percentage of all grant income	Average grant income per organization
Oligarchy	$18.52	14.13	0.38
Representative	$4.73	3.61	0.79
Limited democracy	$106.29	81.08	6.64
Democracy	$1.55	1.18	0.17

organizations with limited democracy. With the Nature Conservancy excluded, the grant income of limited-democracy organizations drops to $9.29 million (27 percent of the total) and the average grant per organization drops from $6.64 million to $620,000. Oligarchic organizations then have over 54 percent of foundation funding.

Locus of Organizational Control

By placing individuals on a movement organization's board of directors and by making grants conditional, a foundation can significantly restrict the organization's power to act independently. This is evident from the results of a 1992 survey in which 265 leaders of environmental organizations were asked to either agree or disagree with a series of statements about foundations and environmental organizations (Snow 1992).

Fifty-nine percent of the leaders agreed that foundations were "unresponsive to grantees' needs" (table 10.20). Only 26 percent agreed that foundations are "blind to their power over grantees." The power of foundations to steer environmental organizations was also evident in interviews that Snow conducted. In discussing the impacts of foundations on environmental organizations, he found that "in the privacy of the interviews, many staff leaders were critical of foundations and of the restriction on activities that comes with heavy reliance on 'soft' funding," and that "some conservation leaders would reject foundation philanthropy entirely if they could figure out a way for their organizations to live without it" (Snow 1992.: 89–90).

The largest influence on environmental social movements comes from the combined efforts of the Environmental Grantmakers Association, which

Table 10.20
Attitudes of environmental groups' leaders toward private foundations.

Statement	Percentage agreeing or strongly agreeing
Foundations should give more funds to general support.	88
Foundations give too little money to local groups.	60
Foundations' officers are unresponsive to grantees' needs.	59
Foundations are blind to their power over grantees.	26

Cockburn and St. Clair (1994: 761) characterize as follows: "Like the oil monopolies of old, the big Eastern foundations that now run the environmental movement don't act separately. Instead, they pool resources under the auspices of the Environmental Grantmakers Association."

The attitude and the intent of some members of the Environmental Grantmakers Association are evident in the transcripts of a 1992 meeting (EGA 1992). During a discussion of environmental advocacy organizations' lack of a unified approach, the following conversation took place (ibid.: 14):

Anne Fitzgerald [audience member]:   Do you detect, though, a resistance in the larger organizations to becoming grant driven?

Donald Ross [director of Rockefeller Family Fund and head of EGA]:   Yeah. I think a lot of them resist.

Chuck Clusen [of American Conservation Association]:   A number of use have been involved in this, Ann. Yeah. There's definitely a feeling on the part of the not for profit organizations that in cases of some of the campaigns like the Ancient Forests Campaign that they resent funders, not just picking the issues, but also being directive in the sense of the kind of campaign, the strategy, the style, and so on. I guess, coming out of the advocacy world, and having spent most of my career doing it, I look at it as, if they're not going to do it on their own, thank God funders are forcing them to start doing it.

Later in the same session, Ross explained his position clearly:

I think that there are things that could be done. I think funders have a major role to play. And I know there are resentments in the community toward funders doing that. And, too bad. We're players, they're players. But I think we touched on a lot of problems, the internal problems within these big groups. the warring factions within them who are all trying to get resources, and there's too many groups and too few resources and all that. I think the fundamental effort that has to be made is a reorganization of the movement. . . . I think we have to begin to look much more at a task force approach on major issues that is able to pool. And the funders can drive that. And part of the reason these groups have been resistant to work with each other is precisely because they want the credit, they want the name, so they can get more funding, either from us—from foundations—or from members. (ibid.: 15)

Another example of foundations' exercising significant control over environmental organizations is found in the activities of the Pew Foundation, which Salisbury (1996: 1) maintains has become "the most active private funder of environmental and conservation organizations in the country." The Pew Foundation is an active player in ecological politics, forming and directing organizations to carry out its issue agenda. "Over the last several years," Salisbury writes (ibid.), "Pew has created and funded dozens of programs and independent organizations to carry out agendas determined by the foundation and its consultants. It has promoted its own causes, pursued its own initiatives, bankrolled its own research, and imposed its own order. Pew is now a player—a 'social investor' and 'civic entrepreneur' in the words of its president."

The Pew Foundation has drawn criticism for diverting money and attention from established organizations, and for undermining grassroots organizations (Salisbury 1996: 26). Critics point to a 1995 episode in which the Pew Foundation withdrew its financial support from a coalition of organizations that had been formed to defend the Endangered Species Act as a result of a dispute over which organization would handle a media campaign. The coalition, which had no further source of funds, was badly split and fell into disarray. Sam Hitt of the Forest Guardians commented: "In effect Pew has used its money like a blunt instrument to destroy a web of relationships that is finally working to save the Endangered Species Act from political death. The major issue here is power and control." (ibid.: 27)

Change of Organizational Strategies
In general, foundation funding influences an organization to move away from confrontation and protest and toward noncontroversial positions and nonconfrontational practices. It also fosters corporatistic decision making (Walker 1991: 108–121). The forest activist Tim Hermack has stated that "foundation money behind a compromise position tempts nonprofits to moderate their hardline stance or risk being left out of the coalition" (Dowie

1995: 52). There is some empirical evidence to support this assertion. As I have noted, the vast majority of environmental organizations use "insider" strategies. Only a few organizations (primarily Deep Ecology and Environmental Justice organizations) use protest strategies, and these organizations receive virtually no foundation support. The overwhelming preponderance of foundation funding goes to organizations that use insider strategies.

In addition to this statistical evidence, there are well-documented cases of environmental organizations' being forced to shift their tactics by pressure from foundations. One such case involved the Natural Resources Defense Council, which in the late 1960s made aggressive use of lawsuits to compel corporations to comply with environmental laws. After complaints from the NRDC's foundation sponsors, the more aggressive lawyers were fired and a board was set up to screen future lawsuits. The board was controlled by the Ford Foundation (Dowie 1995: 37–38).

According to the journalists Silverstein, Cockburn, and St. Clair (1995: 3), "one of the conditions attached to a Pew grant is that the recipient's attention be focused on government actions; corporate wrongdoers are not to be challenged. Thus, with Pew money rolling its way, the environmental opposition becomes judicious, muted and then disappears."

Foundation funding ties a movement organization into a network of financial patrons.[5] This shifts control from the members to the financial patrons. Since foundations work together, this compels a movement organization to either join this network of foundations or do without foundation funding. Summarizing his analysis of foundations' influence on social-movement organizations, Jenkins (1989: 311) maintains that "foundation patronage . . . could be interpreted as creating a neocorporatist system of political representation in which elites exert increasing control over the representation of social interests."

Neocorporatist practices within the environmental movement are well documented. Gottlieb (1993: 117–161) describes in detail the "Group of 10" meetings of the 1980s, in which the chief executive officers of ten of the largest and most influential national environmental organizations convened with large corporate and foundation funders to develop a coordinated strategy. Another example of neocorporatist negotiations is the involvement of environmental organizations with the environmental-mediation movement (Manes 1990: 60). Environmental mediation involves holding meetings of involved parties—such as corporations, government agencies, and environmental organizations—to resolve policy differences. These meetings usually take place under the aegis of a professional mediation service, such as the Keystone Center. Amy (1987: 165–172) has argued that this technique is based on an assumption that core of environmental conflict is rooted in a misunderstanding of the situation, not in fundamental conflicts of interest. Seeing this technique as primarily a process of co-optation, Amy recommends only very limited and cautious involvement by environmental organizations (ibid.: 224).

Reduction of Grassroots Mobilization Efforts

Another effect of foundation funding is the displacement of grassroots mobilization efforts and membership funding by grant-application activities. Furthermore, foundation funding can lead to the decline of efforts to mobilize grassroots support by relieving movement organizations of the continual need to mobilize members and to obtain their financial support. One environmental activist learned about this process when he had to restart his organizing activities after foundation funding for his activities stopped:

> Foundation support is like a drug. You get that check in the mail and you are on the ceiling for the rest of the day, but then it goes away. We became dependent on foundations, and we failed to build a base of support in our communities. Now that foundation support is shifting and in some cases drying up, we're going back to the communities. (Sam Hitt, quoted in Green 1995: 13–14)

There are also less direct ways in which foundation funding can reduce the political strength of social-movement organizations. Jenkins (1987: 314) observed that "the political power of the nonprofits depends largely on their political context, especially the existence of mass movements and supportive elites," and that "although professional advocacy represents increased formal access, if this comes at the cost of greater organizational dependence on elites and diminished mass support, then the gain is probably negative."

Loss of Legitimacy

The legitimacy of an environmental organization as a representative of citizens' concerns can be adversely affected if foundation support leads to professionalization of the organization's staff and to isolation of its directors from its membership. Further, elitism may disengage the leaders of the organization from the members' concerns (Gottlieb and Ingram 1988), thus diminishing their legitimacy in the eyes of the membership. In addition, professionalization of environmental organizations decreases their legitimacy in the policy process. As Jenkins (1989: 312) notes, "Congress is frequently skeptical of environmental movement organizations, viewing them as staff driven and lacking genuine member support."

Co-optation and Ineffectualization
Regardless of how effective mass-mail fundraising and foundation grants may be in the short run, they certainly influence how a social-movement organization is organized and how it carries out its activities. In the long run, they can cause a movement organization to be co-opted or to become ineffectual.

Foundation funding and mass-mail fundraising creates and maintains closed, nonparticipatory "Astroturf" movement organizations based in economic sponsorship. The policies adopted and advocated by an Astroturf organization are developed by the staff and the board of directors under the influence and scrutiny of their foundation sponsors. This compromises the organization's independence, limits its capacity for political mobilization, and subordinates it to the control of the market and the bureaucratic state. How such organizations can bring about meaningful change in a social order while being channeled by the established institutions of economic power is highly problematic.

Characteristics of Environmental organizations Based in Various Discourses

Wildlife Management
Organizations based in the discourse of Wildlife Management are generally very large in economic resources and in membership. They receive most of their income from public contributions and from a number of membership services, little of it from foundations. They are mostly representative or democratic in organizational structure. Education and other nonconfrontational means are their primary strategies for pursuing their goals.

Their issue focus is almost exclusively on utilitarian use of nature for recreational hunting. This severely limits the range of issues and concerns they address. Although their focus is narrow, these organizations provide an instructive example of large, member-funded, participatory environmental organizations that authentically represent their members' concerns.

Conservation

Organizations based in the discourse of Conservation are moderate in size, in membership, and in economic resources. They get much of their income from public contributions and membership services, a moderate amount from foundations. In organizational structure they are mixed, with a slight trend toward more participatory structures. Their strategies for social change are exclusively traditional and nonconfrontational. Although this discourse seeks to provide a universal approach, it has never been able to move beyond a strictly utilitarian focus.

Preservation

Organizations based in the discourse of Preservation are, with the exception of the Nature Conservancy, moderate to large in membership and in economic resources. As was mentioned above, the Nature Conservancy dwarfs all other environmental organizations in terms of economic resources. These organizations, especially the Nature Conservancy, receive the vast majority of foundation support. Nearly two-thirds of these organizations are oligarchic in structure, and as a result these organizations have a strong tendency to become isolated from the members they supposedly represent. Because they receive significant economic support from foundations, these organizations have a substantial potential to be co-opted. All the Preservationist organizations follow insider strategies. Their focus on the protection of wilderness and wildlife limits the range of issues and concerns addressed by the organizations based in this discourse.

Reform Environmentalism

Organizations based in the discourse of Reform Environmentalism are similar to Conservation organizations in number of members and in economic resources. They also receive a large amount of foundation support. More than two-thirds of them are oligarchic. Because they receive significant economic support from foundations, these organizations have a substantial potential to be co-opted. With only a few exceptions, they follow a mix of insider strategies for social change. Owing to its top-down, technocratic approach, this discourse has been unable to develop a cogent political approach or to motivate large-scale political action.

Deep Ecology

Organizations based in the discourse of Deep Ecology are small in number of members and in economic resources. They receive a substantial portion of their income from foundations. Although they also use other strategies, the vast majority of these organizations use unconventional means (primarily protest) to influence the policy process. All of the Deep Ecology organizations examined were oligarchic. Deep Ecology is a limited discourse with few means of addressing social and political issues.

Environmental Justice

Organizations based in the discourse of Environmental Justice are small in economic resources and in number of members. They receive some support from foundations and substantial funding from government grants. Their primary strategy is parliamentarian. In organizational structure these organizations are roughly evenly divided between participatory and oligarchic. The issue focus of these organizations is primarily on the inequitable distribution of effects of ecological degradation to economically disadvantaged or ethnic-minority communities. This issue focus narrows the political scope of their actions and limits their appeal to specific communities.

Ecofeminism

The small size of the sample precludes any real analysis of the economic structure of organizations based in the discourse of Ecofeminism. In general, they appear to be very small organizations, with few economic resources, few members, and virtually no foundation support. Their primary strategy is parliamentarian. The small sample tends toward oligarchic structures, but this cannot be considered conclusive. Though this discourse is very new, it appears to be applicable to a wide variety of ecological concerns. The ability of this discourse to serve as a model for a widespread social movement has not yet been developed.

Diversity and Unity in the U.S. Environmental Movement

The current U.S. environmental movement is a confusing patchwork of overlapping discourses and organizations that, over 150 years, have developed into communities with disparate goals, issue focuses, and organizational structures. Despite attempts by conservationists and reform environmentalists, no master frame has emerged that can unite all these factions into a cohesive political movement.

One way of understanding the diversity of the movement is in terms of distinct components of a potential master environmental discourse. Viewed through the lens of Habermas's theory of communicative action, each discourse makes a partial contribution toward a more comprehensive discourse which could inform the creation of an ecologically sustainable society. As Habermas shows, a discourse must sustain its validity claims to being truthful, morally correct, and authentic. By satisfying these validity requirements, communicative action creates and maintains social action. Habermas's student Klaus Eder (1996: 162–212) argues that an ecological master frame would have to address ecological issues through frames that provide empirical objectivity, moral responsibility, and aesthetic judgment. By creating an environmental discourse that would address all three of these dimensions, the master frame could enable a comprehensive ecological discourse. Specifically, from Eder's perspective, the various environmental discourses form different and partial aspects of the social rationality necessary to justify actions to create an ecologically sustainable society. The discourses of Conservation, Wildlife Management, and Reform Environmentalism comprise the empirical analyses of ecological conditions and their impacts. However, they rely on a utilitarian and anthropocentric morality, and they have virtually no aesthetic dimension. The discourses of Ecotheology, Ecofeminism, and Environmental Justice provide strong moral cases for action, yet they are generally devoid of aesthetic representations of the type of world they value. The discourses of Preservation and Deep Ecology provide extraordinary aesthetic arguments for the protection of the natural world, yet those arguments remain focused on individuals and their personal relationship to nature. Thus, Preservation and Deep Ecology have not provided a wider vision of an ecologically sustainable society.

There are different forms of reasoning about the value of nature. Science can provide information about practical effects on the natural environment. However, this approach does not provide moral or aesthetic reasoning. We also need to develop and integrate cogent moral arguments about the need to protect human health and biodiversity. These are not just technical concerns; they are also concerns that define the good and moral life. We also need to develop aesthetic arguments for the preservation of nature. Aesthetics defines what is the essence and meaning of our existence. Nature certainly takes a central place in our existence. We need all three forms of argument to develop a rational and comprehensive discourse that can be the basis for the creation of an environmental movement capable of reversing the ecological destruction of our world.

11

Agency, Democracy, and the Environment

If there is any small remnant of utopia that I've preserved, then it is surely the idea that democracy—and the public struggle for its best form—is capable of hacking through the Gordian knots of otherwise insoluble problems. I'm not saying that we're going to succeed in this; we don't even know whether success is possible. But because we don't know, we at least have to try.
—Jürgen Habermas, 1991[1]

It was a wonderful early spring day in Washington. The annual board meeting was being held at an exclusive Georgetown mansion. A few minutes before the meeting, the members of the board began to arrive in their chauffeured Mercedes limousines. One by one, they were dropped off at the grand entrance to the mansion. This was no ordinary board of directors. On it were a former director of the National Coal Corporation of America, a managing director of BancAmerica, and a number of prominent private investors (Issel 1999). As they entered the mansion, a small group of protesters were there to meet them. They were protesting several of the organization's activities, including collaboration with the American Chemical Association in developing a chemical toxicity testing program based on extensive animal testing, support of the North American Free Trade Agreement, advocacy of bailouts for nuclear utilities, and delaying action on global warming.

This protest would not be unusual, except for the fact that this was not a meeting of the board of directors of a multinational corporation. Rather, it was the annual board meeting of the Environmental Defense Fund. The EDF started out in 1967 as a ragtag, informal group of lawyers and biologists that filed a legal suit to halt spraying of DDT on Long Island. Now, 33 years later, this group was itself the subject of protest from another group of environmentalists.

Late in 1996 one of the most respected U.S. environmental organizations, Environmental Action, closed its doors. This organization was born as the organizing committee of the first Earth Day in 1970. It was an original champion of environmental justice. Its "Dirty Dozen" campaign was highly effective in generating criticism of public officials who were responsible for environmental degradation. In addition, it had a steady and loyal membership. Despite all these factors, it was unable to obtain sufficient funding to keep going (Bendavid 1996). In 1997, Greenpeace, another stalwart environmental organization, laid off several of its workers and closed numerous offices around the country. The decline of these two organizations and the demonstration against the Environmental Defense Fund are symptomatic of the current state of the environmental movement in the United States.

In his 1995 book *Losing Ground*, Mark Dowie argued that the U.S. environmental movement had become ineffective. One reason he cited was the co-optation of the mainstream environmental groups under the rubric of "third-wave environmentalism," which began in the Reagan era:

The mainstream movement responded to Reagan by forming the harmless and stubbornly elitist Group of 10, creating its own irrelevance by remaining middle class and white, pursuing "designer issues" expedient for fundraising, focusing on Washington, lobbying the wrong committees, failing to move women and minorities into top jobs, building ephemeral memberships with direct mail, ignoring the voice of vast constituencies, and eventually—under the rubric of third-wave environmentalism—cozying up to America's worst environmental violators. (Dowie 1995: xiv)

Two responses to the development of bureaucratically structured oligarchic organizations were the formation (in the mid 1980s) of a number of new movement organizations based on the discourses of Deep Ecology, Environmental Justice, and Ecofeminism and the emergence (in the 1990s) of ecotheology (Wilson 1990: 71; Gottlieb 1993: 3–7; Bullard 1993: 23).

It is clear from the writings of its founders (Foreman 1991: 16–23; Manes 1990: 57–65) that Earth First! was founded in response to the perceived co-optation of professional environmental groups. Paul Watson, the founder of Sea Shepherd Conservation Society and one of the founders of

Greenpeace, succinctly expressed another example of this attitude. Watson left Greenpeace because it was getting too bureaucratic for his taste. In summarizing this position, Gottlieb maintains that the professionalized groups were not in touch with the needs of grassroots groups. This led to the creation of new environmental groups among the populations whose interests were not being represented by the mainstream environmental organizations (Gottlieb 1993: 185–188).

Serious splits between mainstream and radical groups have been caused by divergences in strategies and tactics and by basic class and racial animosities. Commenting on the lack of healthy interaction between radical and moderate environmental groups, the chairman of the Sierra Club noted that "the dilemma is how to get these two ingredients into a productive relationship," and that "apart, the radical groups may expend their energy with little tangible results, whereas the mainstream groups may lose their way with no clear vision to pursue" (McCloskey 1992: 85).

Yet, despite the development of these new discourses and social-movement organizations, the political strength of the environmental movement has not increased significantly. None of the new organizations has had anywhere near the type of expansion that occurred in the 1960s. This lack of a mass public movement for ecological concerns has been noted by Killingsworth and Palmer (1992: 7):

What is true for the government-sponsored social engineers that grew up in the conservation movement at the beginning of [the twentieth] century is true also for newer groups like the deep ecologists, wilderness preservationists, ecoanarchists, and green politicians: They have been unable to create strong communicative links with the mass public, links that would support a strong power base for reformative actions.

The result is a polyglot of groups, lacking any coherent strategy, badly fragmented, and consequently unable to muster sufficient political power to realize their goals.

It is clear that we need to find new ways of dealing with the ongoing process of ecological degradation. The many well-intentioned efforts have not proved adequate to our historical situation. We have to make a dramatic increase in the rate of social change toward a sustainable society if we are to avert the worst of the projected consequences of ecological degradation.

Distorted Communication in the Environmental Movement

The core argument of this book is that the restriction of communicative action restricts the learning capacity of modern society. Without this self-corrective capacity, the destruction of the ecological basis of our social order cannot be adequately addressed.

Environmental organizations play a crucial role in restoring this self-correction capability through the creation of a robust civil society and restoration of the public sphere. A restored and active civil society would increase the learning capacity of the social order and foster the initiation of its restructuring toward ecological sustainability. I have identified three things that limit the ability of these organizations to accomplish this task: failure to develop a comprehensive ecological discursive frame, co-optation and steering by external funding organizations, and the creation of oligarchic movement organizations.

The Failure of Ecological Discourses

The analysis of discursive frames has shown how we can enter into discourse traps and ideologies that dichotomize the causes of ecological degradation, limit and channel debate into ideological contests, and restrict the nature of the options that can be considered in the search for solutions (Killingsworth and Palmer 1992: 9–10). A cultural critique of the limitations of the discourses can assist us in identifying these characteristics in the different ecological discourses and in attempting to avoid these traps.

There is no ecological discourse that can currently unite the disparate discourses of the U.S. environmental movement. As Harre et al. note (1999: 20), environmentalism has "become a worldwide cluster of dialects" that, as yet, is "far from the expression of a unified voice." Instead, there are multiple and partial discourses that are unable to appeal to a wide audience.

One reason for this lack of effectiveness is the lack of a common notion of what would constitute the good life. Taylor (1992: 137) notes that "the story of American environmental theory in the twentieth century is primarily the story of an increasingly obscured political vision." In fact, the various segments of the environmental movement do not even engage one another to attempt to form such a community. Therefore, these movements are fragmented, without a common vision or purpose (Bruner and

Oelschlaeger 1994). The individual concerned citizen is left to his or her own devices to make sense out of this fragmented dialogue in which each group talks past the other in a process of mutual incomprehension and suspicion. This leaves consciousness unable to synthesize the current situation, and fragmented in a series of unrelated ecological discourses.[2] This confusion limits the mobilization of citizens into taking political action (Habermas 1987a: 355). As a result, "what is missing from American environmental policy today is a coherent vision of the common environmental good that is sufficiently compelling to generate sustained public support for government action to achieve it" (Andrews 1999: 370).

In addition, the existing discourses block or mask the social origins of ecological degradation. The discourses of Conservation and Reform Environmentalism see scientific management of the natural environment as the solution. The practice of "ecological modernization" accepts the basic structure of the social order and seeks to maintain it through technocratic management. Despite more than a century of the application of conservation, "careful, scientific management of nature has failed to fully understand and protect the natural world in all its complexity and fragility" (Taylor 1992: xi). Furthermore, these discourses effectively work to maintain the existing social structures. Scientific discourse attempts to avoid political discussion in its descriptions of processes of ecological degradation. Killingsworth and Palmer (1992: 19) note that, by adopting this approach, "the formal discourse of mainstream science preserves its 'contemplative' distance through careful restrictions on language and genre and thereby remains ironical on political questions, creating a kind of apolitical politics out of its vaunted neutrality." By creating a technocratic value-neutral discourse, it obscures consideration and discussion of the social causes of ecological degradation. It also removes moral considerations from public policy discussions. Finally, it serves as an ideology that limits the participation of citizens in ecological discussions. If scientific language is taken to be the lingua franca of ecological issues, then those citizens who do not have credentials in the appropriate sciences are not legitimate participants in the dialogue. Without the power to speak the specialized discourse of science, they are without a voice in this public debate. Instead, the citizens are reduced to the status of a population to be managed, and to be properly educated through techniques of "risk communication." Habermas (1970: 68) has described this process as the scientization of politics and has argued that it "reduces the process of democratic decision-making to a regulated acclamation procedure for elites alternately appointed to exercise power." As a consequence of this scientization, neither the discourse of Conservation nor that of Reform Environmentalism has been able to "maintain and develop this relationship between our political values and the ways in which we do or should value the environment" (Taylor 1992: xii–xiii). Since these discourses are unable to develop a dialogue that integrates scientific analysis of ecological problems into a notion of a good and moral society, they have had only a limited appeal among technocratic managers. Their lack of concern for social and political justice not only limits their appeal; it also serves to preserve the current social order (Brown 1998).

Co-optation of Environmental Organizations

As was discussed extensively in chapter 10, external funding draws an environmental organization into an existing network of economic and political power. Its participation in this network is contingent on shifts in its behavior, including the development of a centralized and professional staff, the undertaking of projects suggested by the funding agencies, the development of moderate political goals, and the adoption of conventional political tactics. This limits the range of organizational forms and goals, shifts the locus of control of the organization from the membership to the financial patrons, reduces the organization's efforts at grassroots mobilization, and limits the influence of the membership in deciding the direction of the organization. Thus, instead of being governed by democratic and rational discourse, movement organizations become subordinated and controlled by market institutions.

Creation of Oligarchic Environmental Organizations

The third process that limits the social learning capacity of environmental organizations is the creation and maintenance of movement organizations that are oligarchic in structure. Oligarchy blocks citizen involvement, limits the range and scope of alternatives considered, and limits the organization's capacity for mobilization.

There are several processes that foster this type of organizational structure. The first, discussed above, is the influence of foundation funding, which requires a centralized management and accounting system and thus concentrates power in the central staff of the organization. The second is professionalization. The insider strategy involves many environmental organizations in the legal and technocratic policy process and thus requires them to develop professional staffs and bureaucratic structures. This also contributes to the creation of an organizational structure in which members have little room for participation. The third process is the adoption of mass mailing as a fund-raising technique, which transforms members into contributors and replaces face-to-face interaction with anonymous and transitory mail contact.

It is clear that some discursive frames foster the development of oligarchic organizations. As I showed in chapter 10, two-thirds of the organizations based in the discourses of Preservation and Reform Environmentalism and all of those based in Deep Ecology are oligarchic. In the case of Reform Environmentalism, it is easy to see why. Reform Environmentalism sees the solution to environmental problems in top-down system management (Taylor 1992; Evernden 1992a). Scientists, because of their expert role, are seen as having a prominent role in the "scientific management of the environment" (Taylor 1992: 50). Citizens and "decision makers" play only bit parts, heeding the advice of the scientists. There is no identified need to involve the public, except to obtain financial support. In the case of Preservation and Deep Ecology, these oligarchic tendencies are most probably based on the nature of the discourses, which emphasize individual insights and aesthetic knowledge. Control of these organizations is justified not by democratic debate but by access to unique aesthetic insights. Such a legitimization of power thus enables the creation of oligarchic organizations.

Limiting Social Learning
The three processes described above contribute to the restriction of an environmental organization's effectiveness in communicating the concerns of its members from their lifeworld into the public sphere. If the communicative basis of social order in the lifeworld is systematically blocked from correcting social organizations, the ability of society to adapt to changing circumstances is limited.

American society's capability for self-correction is systematically blocked by the institutions of capitalism and the bureaucratic state. The progressive deterioration of communicative action has created a society that is unable to renew its social institutions to adapt to changing circumstances. The failure of ecological discourse to provide a challenge to this situation limits its ability to mobilize effective political power. In addition, there are powerful social forces, in the form of foundation steering and the creation of oligarchic movement organizations that limit the social learning capability of our society. These social factors result in the continued degradation of the natural environment.

Social Learning for an Ecologically Sustainable Society

To begin addressing the social origins of ecological degradation, society will have to expand its capacity for social learning. Social learning depends on the creation of alternative worldviews, the open communication of these realities into the public sphere, and the use of this knowledge to develop pragmatic and incremental social experiments. An increase in social learning depends on the creation of social institutions able to translate the impulses of the lifeworld into effective political discourse in the public sphere, thereby subordinating the institutions of the market and the bureaucratic state to democratic control.

One means through which social learning can occur is institutional change fostered by social movements. My research identifies three areas in which action can be taken to begin to address the current situation: development of an ecological metanarrative, reform of the practices of external organizations, and encouragement of the formation of democratic environmental-movement organizations.

Toward an Environmental Metanarrative

The serious divisions that exist within the environmental movement require a restructuring of the political and ideological cleavages that exist within and among the various environmentalisms so that a coherent project of collective action can be developed (Eder 1996a: 205; 1996b: 164). To overcome these divisions, we need to create a community that can unite around the goal of creating a democratic and ecologically sustainable society. To do this, we need to create some form of common discourse. Torgerson (1999: 20) argues that we need to develop "an ecologically informed discourse that . . . challenges the monological administrative mind and the prevailing discourse of industrialism."

Arguing that this discourse takes the form of a metanarrative or a masterframe, Eder (1996a: 207) asserts that an environmental masterframe "identifies every element in public discourse on the environment: to ethical questions, scientific theories, and to literary expressions of the relationship of man with nature."[3] He distinguishes a number of environmental subframes, such as Deep Ecology and Ecofeminism (ibid.: 203–214). Thus, a masterframe is a collection of a number of cognitive, ethical, and aesthetic arguments for the preservation of nature.

An example of a partial metanarrative of nature is represented by the term *biodiversity*. This discursive invention unified disparate discourses and groups that were concerned about destruction of the natural environment due to deforestation, overfishing, introduction of exotic species, hunting, habitat destruction, and extinction of species. It did not absorb or destroy the other discourses. Rather, it expanded the concerns of the various groups to see their common purposes. By so doing, it lead to an increase in collective action. A similar environmental metanarrative would allow for "the construction of a new common sense which changes the identity of the different groups so that their differing practices are able to complement one another (Torgerson 1999: 47). There is no one particular environmental discourse that could function as a metanarrative. The development of the various environmental frames has resulted in a number of particular discourses which are unique cultural responses to specific conditions of ecological degradation (Eder 1996b: 163). Instead, an environmental metanarrative will consist of multiple forms of arguments to motivate action in different social orders. There is no requirement that joint action be based on one set of cultural beliefs; the only requirement is that there exist good reasons to act in a particular manner. "Political unity," Schlosberg (1998: 87–101) notes, "does not require that a political agreement be reached based on identical reasons. Rather, unity can be achieved through recognition and inclusion of multiplicity and particularity."

The theory of communicative action can inform the creation of a metanarrative by describing both the types of arguments the metanarrative would have to make and the social conditions under which it would be created. First, the metanarrative must provide aesthetic, moral, and cognitive reasons for collective action. In order to create a rational agreement about what joint action should be followed, a discourse must establish that it accurately represents the objective world, that the acceptance of a proposed action is in accordance with other existing cultural norms and beliefs, and that the statement is an authentic representation of the speaker's inner self.[4] Hence, the development of collective action depends on a discourse's sustaining the validity claims of truth, normative rightness, and authenticity. This means that the multiple and partial discourses on the natural environment must be integrated to form a coherent discourse that can provide cognitive scientific, normative, and aesthetic rationales for the preservation of nature (Eder 1996a: 215). Thus, contributions are needed from all the environmental discourses. The existing environmental discourses form the starting point for such an effort. According to Killingsworth and Palmer (1992: 266), an ecological metanarrative "will . . . draw energy and direction from them and in turn will influence their sense of purpose and their understanding of their relationships to other discourses. The continuous narrative of an environmentalist culture will, above all, be the medium through which communicative action is realized and perpetuated." Thus, to be successful, an environmental metanarrative must recognize the validity of the great variety of viewpoints and rationales with regard to protecting the natural environment, and this knowledge must inform collective action (Schlosberg 1998: 21). Thus, we will need to "enlarge the range of voices in our conversation and with them the means of considering our relationship with the natural world" (Killingsworth and Palmer 1992: 79).

An ecological metanarrative would draw on the special management, scientific and legal capacities of the Conservation, Wildlife Management, and Reform Environmentalist discourses to ensure scientific competence and adequately address scientific questions. To develop new normative criteria, it would need to encompass the moral fervor and commitment of the ecotheologists and the deep concerns over equity and justice of both the Ecofeminist discourse and the Environmental Justice discourse. To address our images of what constitutes the good life, it would need to incorporate the aesthetic insights provided by the discourses of Preservation and Deep Ecology.

The theory of communicative action also defines the process through which this metanarrative would be created. Here Habermas's communicative ethics specifies that the process of validating a discourse requires an open speech community in which the unforced force of the better argument prevails. This ethical relationship, a presupposition of mutual communications, requires basic recognition and acceptance of others and respect for the autonomy and integrity of the others' identity and selfhood (Schlosberg 1998: 68). Thus, the theory of communicative action specifies that the creation of an ecological metanarrative must occur through open communication based on solidarity and mutual respect.

An open dialogue among all environmental organizations would create a democratic community, which could then debate and develop coordinated actions to deal with ecological degradation. Torgerson (1999: 160) has labeled the arena in which this dialogue would take place the "green" political sphere. It is not a specific institution or think tank. Rather, it defines a change in the relationship of the various environmental communities that would enable them to engage one another in a community of dialogue (ibid.: 161). The "green" public sphere would define a space in which "industrialist presuppositions do not prevail" (ibid.: 162). This public space would be the arena in which environmental politics would take place and meaningful disagreements and debates about our society and the actions necessary to create an ecologically sustainable society would be carried out. Creating an environmental metanarrative through this dialogue would enable the creation of an environmental community capable of democratically discussing proposed actions and then acting together (ibid.: 107).

Reform of External Organizations

The role of external funding of movement organizations also needs to be significantly reformed. Although one would hope that this analysis would lead some foundations to examine the impacts of their practices, self-reform of that kind is unlikely. What would be required to change the practices of these organizations would be full disclosure of grant information and limits on the life spans of foundations. First, the Internal Revenue code should be reformed to require all nonprofit organizations to fully report all grants, contracts, and other sources of income, including economic resources received from foundations, corporations, and government agencies as well as substantial individual contributions. This would allow monitoring of what economic forces a social-movement organization is being subjected to. Second, the laws regarding corporate and foundation funding of non-profit organizations should be reformed. Plutocratic influence on social movements is fundamentally undemocratic and should end. At a minimum, philanthropic endowments should not be allowed to exist in perpetuity. A reasonable time period for the life of a foundation should be established. In addition, the boards of private foundations and corporations should be revised to become more representative of the public (Ostrander 1995; Colwell 1993: 205).

Democratic Environmental Organizations

To serve as agents of social learning, environmental organizations themselves must have a democratic structure. By communicating the imperatives of the lifeworld to the public sphere, environmental organizations can serve as a bridge between the individual's lifeworld and the larger social order. This would allow for the possibility of increasing the society's learning capacity, and it would foster the initiation of a restructuring of the social order toward ecological sustainability. It would also enhance the mobilization, the legitimacy, and the political power of these groups.

What is needed is the development of creative ways to expand members' participation in environmental organizations while preserving the organizations' technical and scientific expertise. There is no set formula for how this increase in participation would be accomplished, nor is there a need for direct democratic participation in the organization. "It is," Habermas notes (1979: 186), "a question of finding arrangements which can ground the presumption that the basic institutions of the society and the basic political decisions would meet with the unforced agreement of all of these involved, if they could participate, as free and equal, in discursive will-formation. Democratization cannot mean an a priori preference for a specific type of organization, for example for so-called direct democracy." To realize this objective, environmental organizations will have to engage in practical experimentation. My research suggests that social experimentation in two areas might prove useful.

First, the influence of foundation funding on the structure of environmental organizations should be addressed. Foundation funding can severely marginalize or restrict the participation of members in an organization. This process should be consciously taken into account, and steps should be taken to ensure that foundation support does not lead to a deterioration of members' participation.

Second, the potential for developing authentic membership involvement in governing nationwide environmental organizations should be explored. There are large, democratic environmental organizations. The Sierra Club is one. Its means of citizen involvement should be examined and copied to further expand the participatory structure of environmental organizations. Slick, commercial-style, mass-mailing fund solicitation should be replaced by authentic participation. One step in this direction would be to replace the fake surveys mailed out to encourage membership renewals with real surveys of members. Organizations' publications could air different points of view and criticisms. To participate meaningfully, members need information about an organization's activities and a means of communicating with the leadership. As Rittner (1992) noted, a vast array of information on environmental organizations has been made available on the Internet. Environmental organizations could enhance the knowledge of their membership by making their finances, sources of support, grants, and policies available for public inspection. In addition, use of the Internet would allow members to provide input directly to the leadership.

Finally, the U.S. government's procedures for treating environmental organizations as representatives should be changed. Advisory boards, regulatory panels, and other groups are often put together by government agencies, purportedly to enable citizens or their representatives to participate in the development of government policies. Many of the national environmental organizations examined above are included in such groups as representatives of the environmental community. My analysis raises questions about this practice. An oligarchic organization does not meet either the theoretical idea of citizen participation as defined by the theory of communicative action or the legal requirements of a representative organization. Yet oligarchic organizations are regularly included in government deliberations on the assumption that they are authentic community representatives.

There are, of course, many other creative ways in which the participatory nature of existing environmental organizations can be enhanced. The ideas for democratic reform contained in this chapter are merely suggestions for beginning a dialogue.

Conclusion

A sustainable society is not a utopia but a very real and practical goal. How we go about getting to that goal will influence the shape of our society. There is no separation of ends and means. What we are trying to do is create a way of life that recognizes and values nature and, at the same time, preserves human dignity and autonomy.

The strength of a democracy is that it allows us to talk to one another to solve our common problems. Through mutual recognition we establish the basis for a truly human and enlightened society. This mutual recognition can create a real community of shared interests and actions toward realization of the common good for ourselves and for all the other beings with which we coexist on this planet.

The increasing environmental degradation makes it appear certain that disruption and conflict regarding the relationship of our society to nature will grow. When any society begins to destroy the physical basis of its existence, a fundamental revision of the social order becomes inevitable, either through environmental disaster or through a negotiated and chosen future. To the extent that this book can expand our self-understanding, it can help us in renegotiating our social order and in choosing our future.

Restructuring our society to be in harmony with the natural environment is a long and difficult task. As Weber noted (1918/1978: 128), politics is "a strong and slow boring of hard boards." The degree of social change required is substantial. The creation of a society that is ecologically sustainable and the creation of a society in which human dignity is preserved are linked. One cannot be realized without the other. To create a positive cultural alternative, we need to invent new ways of envisioning our relationship with nature and with one another, then act to realize our visions. By acting in this manner, we may have a chance to create a more rational and just social order that can address the threats to the global environment, and, at the same time, preserve and expand human agency.

Appendix: Methodology

To carry out the detailed examination of U.S. environmental organizations discussed in chapters 6–10, the following methods were used.

Compiling a List of National Environmental Organizations

The first task was to compile a list of national environmental organizations from which a sample could be drawn. A definition of an environmental social-movement organization using the criteria defined by Zald and Garner (1987) was developed. An environmental organization was defined as an identifiable association of citizens, primarily concerned with improvement of the natural environment, that takes action to attempt to realize some public or general good.

Two key aspects of the definition for purposes of the analysis were that the organization has a membership (or at least that one can join as a contributor) and that the organization acts to create some form of social change. Several types of organizations that are closely related to environmental social-movement organizations, including government or government-sponsored groups, professional associations, scientific organizations, zoos, aquariums, natural history societies, policy institutes, and think tanks, do not meet the second criterion.

The following three distinct data sources were used.

The Internal Revenue Service's List of Nonprofit Environmental Organizations

The first data set was a computerized file of nonprofit organizations derived from records maintained by the Internal Revenue Service. Organizations with the IRS Activity Codes shown in table A.1 were selected.

Table A.1

Code	IRS Category
350	Preservation of Natural Resources (Conservation)
351	Combating or Preventing Pollution (Air, Water, Etc.)
352	Land Acquisition for Preservation
353	Soil or Water Conservation
354	Preservation of Scenic Beauty
355	Wildlife Sanctuary or Refuge
379	Other Conservation, Environmental or Beautification Activity
529	Ecology or Conservation Advocacy
541	Population Control Advocacy

The *Encyclopedia of Associations*
The second data set was the 1995 *Encyclopedia of Associations*. Organizations listed under the headings shown in table A.2 were added to the list.

The National Wildlife Federation's Conservation Directory
The third data set was the 1995 National Wildlife Federation Conservation Directory. Once the initial list had been compiled, a process of reviewing and purging the list based on Brulle 1995a was completed. At the completion of this process, 1151 organizations remained on the list.

Developing a Sample of National Environmental Organizations

Because it was impossible to obtain, verify, and code detailed information on all the environmental organizations, a sample of national environmental organizations was developed through consultation with several knowledgeable individuals in the field of environmental sociology (Brulle 1995a). Because of an interest in obtaining a significant number of environmental organizations in each discursive frame, a special effort was made to include organizations from smaller segments of the U.S. environmental movement with discursive frames of Environmental Justice, Ecofeminism, Deep Ecology, and Ecotheology. The result of this process was the list of 106 environmental organizations shown in table A.3.

Table A.2

Acid Rain
Appropriate Technology
Conservation
Ecology
Environment
Environmental Health
Environmental Law
Environmental Quality
Natural Resources
Parks and Recreation
Pollution Control
Public Lands
Wildlife Conservation

Developing Detailed Data on Each Organization

Detailed information on the organizations in the sample was gathered in four areas.

Discursive Frame

The discourse of each organization was coded into one of the nine discursive frames listed in chapter 4. This coding was based on the process developed in Brulle 1995a. This process yielded an acceptable level of intercoder reliability (Perreault and Leigh 1989: 146).

Annual Income and Expenditures

To obtain information on the annual income and expenditures of each organization, the annual "Return of Organization Exempt From Income Tax" (IRS 990) was reviewed. These forms were obtained by mailing written requests to each organization. When organizations did not supply this form after one original and two follow up written requests, the returns were obtained by written request to the Internal Revenue Service.

In addition, because the Annual Information Return does not distinguish between membership donations and contributions from foundations

Table A.3

Wise Use
Alliance for America Inc.
American Land Rights Association
Blue Ribbon Coalition Inc.
Center for The Defense of Free Enterprise
Citizens for A Sound Economy Inc.
Conservationists With Common Sense
Defenders of Property Rights
Keep America Beautiful Inc.
Multiple Use Association
National Coalition for Public Lands & Natural Resources
National Council for Environmental Balance
Putting People First
People for The West

Conservation
American Farmland Trust
American Forests
Elm Research Institute
Izaak Walton League of America Inc.
National Arbor Day Foundation
Rails To Trails Conservancy
Scenic America
Trust for Public Land

Wildlife Management
Boone & Crockett Club
Ducks Unlimited Incorporated
National Wildlife Federation
Quail Unlimited Inc.
Rocky Mountain Elk Foundation
Trout Unlimited
Whitetails Unlimited

Preservation
Appalachian Mountain Club
Audubon Naturalist Society of the Central Atlantic States
Conservation International Foundation
Defenders of Wildlife
International Wildlife Foundation
Mono Lake Committee

Table A.3 (continued)

National Audubon Society Inc.
National Parks & Conservation Assn.
Nature Conservancy Inc.
North American Bluebird Society
North American Wildlife Foundation Inc.
Save The Redwoods League
Sierra Club
Treepeople Inc.
Wilderness Society
Wildlife Conservation Society
Wildlife Society
World Wildlife Fund Inc.

Reform Environmentalism
Air and Waste Management Association
Alliance for The Cheseapeake
American Littoral Society
American Rivers Inc.
Americans for The Environment Inc.
Center for Marine Conservation Inc.
Chesapeake Bay Foundation Inc.
Clamshell Alliance
Clean Water Action
Cousteau Society Inc.
Earth Island Institute Inc.
Environmental Action Inc.
Environmental Defense Fund Incorporated
Friends of The Earth Inc.
Friends of The Sea Otter
Greenhouse Crisis Foundation
Greenpeace
Institute for Local Self Reliance Inc.
League of Women Voters of The United States
Lighthawk
National Coalition Against The Misuse of Pesticides
National Toxics Campaign
Natural Resources Defense Council Inc.
Population Council Inc.
Population Environment Balance Inc.
Population Institute
Rachel Carson Council Inc.

Table A.3 (continued)

Sierra Club Legal Defense Fund Inc.
Society of Women Environmental Professionals
Union of Concerned Scientists Inc.
Zero Population Growth Inc.

Deep Ecology
Alliance for The Wild Rockies Inc.
Cenozoic Society
Earth First!
Elmwood Institute/Center for Ecoliteracy
Native Forest Council Inc.
Planet-Drum Foundation
Rainforest Action Network
Sea Shepherd Conservation Society

Environmental Justice
American Indian Environmental Council
Association of Forest Service Employees for Environmental Ethics
Citizens Clearinghouse for Hazardous Waste
Government Accountability Project Inc.
Morning Star Foundation Inc.
National Tribal Environmental Council
Native American Fish & Wildlife Society
Native Americans for A Clean Environment
TV Free America

Ecofeminism
Mothers & Others for A Livable Planet
Mothers and Others for Pesticide Limits Inc.
Women In Environment and Development Organization
Women In Natural Resources
Womens Council On Energy and The Environment
World Women In The Environment (Worldwide)

Ecotheology
Christian Environmental Association
Christian Society of The Green Cross
Coalition On The Environment and Jewish Life
Evangelical Environmental Network
Floresta USA
National Religious Partnership for the Environment

or corporations, two further steps were performed to fill obtain further information on these sources of income. First, the Foundation Center's database was used. This data set is known to understate the amount of foundation grants. Second, the annual reports were reviewed to see if contributions from specific corporations were listed. They usually are in different categories—for example, listing the corporations who contributed between $50,000 and $100,000 in any one year. By multiplying the number of contributions from corporations by the minimum amount of the category, which in this case would be $50,000, a rough estimate of corporate contributions to groups that listed this information in their annual report was obtained. This method yields a conservative estimate of the amount of corporate contributions.

Internal Organizational Structure

Information on the internal organizational structure of the organization was obtained by obtaining copies of the by-laws of the organization. When organizations did not supply this form after one original and two follow up written requests, the returns were obtained by written request to the Internal Revenue Service and obtaining a copy of the organization's original "Application for Recognition of Exemption" (IRS 1023). The by-laws that were submitted to the Internal Revenue Service in this organization's (IRS 1023) was used.

Coding of the organization's internal structure was through a two step process. First, data was gathered from a review of the by-laws to determine the degree of democracy in the organization. This was based on the strategy of Devall (1970), who developed a series of criteria to typify the organizational type that characterized the Sierra Club. Based on Michels 1915 and Lipset et al. 1956, this index contains nine variables. These criteria were operationalized by answering the question(s) shown below through a review of the by-laws and literature of the organization.

amount of bureaucracy   *Organizations with large, professional staffs are less democratic than those organizations with smaller staffs.*
Does the organization have a professional staff?

structure of organization   *Organizations with a pluralist, decentralized structure, with the opportunity for multiple centers of power to exist are more democratic than centralized organizations.*
Does the organization have chapters or subgroups?

status of leaders and members   *Organizations with a small amount of status incongruity between the leaders and members of the organization will be more democratic than organizations where this incongruity is high.*

Are there limits to the number of terms that can be served by members of the Board of Directors, or officers of the organization?

Are the members of the Board of Directors or officers of the organization elected?

development of organizational goals   *Organizations where the overall goals and programs of the organization are fluid and changeable are more democratic than organizations with a fixed set of goals, and the decisions implementing them are seen as technical or administrative matters.*

Is there a provision for a referendum to be held in the by-laws?

Are there provisions for members to communicate with the Board of Directors or officers of the organization?

Does the organization's literature contain information on the income and expenses of the organization?

internal conflict   *Organizations with a moderate level of conflict are more democratic than organizations that do not have any defined cleavage and Circulation of Dissenting Opinions—Organizations where members are able to receive dissenting opinions are more democratic than organizations where members cannot receive this information (Combined for coding procedures.)*

Does the organization's literature contain dissenting opinions regarding organizational matters of policy/program initiatives?

characteristics of members   *Organizations with interested and informed members are more democratic than organizations with apathetic members.*

Are board meetings required to be public?

Does the organization's literature contain the results of a Board of Director's meeting?

Does the organization's literature contain results of an election?

power distribution   *Organizations with no overwhelming faction are more democratic than an organization with one faction holding a strong majority position in power.*

Are there requirements for some form of proportional representation on the Board of Directors?

Can members nominate individuals to run for election to the Board of Directors or as officers of the organization?

Can members vote for members of the Board of Directors or officers?

participation skills   *Organizations with opportunities for members to learn how to participate in organizational politics outside of the control of*

the official hierarchy are more democratic than organizations where these skills can only be learned by participating within the hierarchy.
Are the rules for elections specified in the by-laws?
Does the organization's literature contain notice of a meeting of the Board of Directors?
Does the organization's literature contain notice of an election?

Based on the results of this series of questions, each organization was assigned to a type of organization using the following criteria.

oligarchy   Governed by Board of Directors. The Board of Directors is a self-replicating mechanism and elects the officers of the organization. No provisions for individual member input exist.

representative   Members can elect representatives for their local chapter of the organization. These representatives then participate in the selection of the board of directors, officers, and policies of the organization.

limited democracy   Governed by mix between Board of Directors and members. Individual members can nominate/elect some of the members of the board of directors or officers of the organization. However, certain aspects of organizational control are specifically delegated to the Board of Directors.

democracy   Governed by members. Board of Directors and Officers of the Organization are nominated and elected by membership. Policies of the organization can be debated and voted upon by individual members.

Strategy for Social Change
Each organization's primary strategy for realizing its objective was coded into one of the five categories (Lofland et al. 1989) through a review of the organization's annual report and copies of its publications. The five categories are as follows:

transcendence   Bring about a rapid shift in consciousness through charismatic and aesthetic demonstrations.

education   Instruction of population in necessary changes through development and dissemination of publications and materials.

parliamentarian   Work through existing legislative and bureaucratic process to create incremental changes.

protest   Force change by noncooperation and/or disruption of regular routines.

prophecy   Bring about deep moral regeneration of society by creating separate communities that rebuild self and society.

Analyzing the Data

This data set was used to examine the distribution of organizational forms within environmental organizations and to test the relationships between organizational form and discourse. This relationship was tested using both Pearson's χ^2 (Blalock 1979: 279–292) and Goodman and Kruskal's τ proportional reduction in error tests (Blalock 1979: 307–310). The results of the statistical tests of the key tables are reported in the notes to chapter 10.

Notes

Chapter 1

1. The Comprehensive Environmental Response, Compensation and Liability Act, commonly known as Superfund, is discussed in chapter 9.

2. This raises concerns about authoritarian solutions to ecological problems and about the possibility of ecofascism. The notion of ecofascism as a viable way to handle the environmental crisis seems is a difficult concept to maintain, however. Fascism is generally defined as involving political authoritarianism with a corporatistic control over markets. Ecofascism thus implies control by the state of both the political and the economic sphere. Creation of an ecofascist a social order would be premised on the central control over investment decisions. This would effectively amount to a demise of the market system. This seems improbable from a practical political viewpoint and in view of the logic of social order (D'Amico 1991). In addition, this position ignores the difficulty of dealing with environmental problems from a centralized technical viewpoint. There is an impressive body of work that details the limits of the technocratic model in dealing with ecological problems. Hence, for practical political reasons and owing to the and limitations of social learning, a centralized decision-making body probably would be unable to deal effectively with worldwide environmental problems. For extended discussions of ecofascism, see Paehlke 1988 and Zimmerman 1997.

3. There has been an extensive debate between critical theory and the other major schools of thought. For summaries of the epistemological arguments on which the theory of communicative action is based, see Radnitzky 1973, Bleicher 1982, and Holub 1991. The communicative ethics developed by Habermas has been the subject of intense debate. For a discussion of the viability of this ethical position, see Benhabib and Dallmayr 1990.

4. There are several important exceptions, including the work of Webler (1992, 1994), Fisher (1990), and Drysek (1987, 1995).

5. For a summary of the debates surrounding the work of Habermas, see Holub 1991.

6. Throughout the book, I use "environmental movement" and "environmental-ism" as umbrella terms. Reform Environmentalism, which I discuss in chapter 8, is a particular historical discourse. It defines a certain sector of these social move-ments. However, it is not the only discursive frame. Only its particular historical position of cultural dominance leads to its current identification as a general term that supposedly can encompass the great variety of discourses described in this book.

Chapter 2

1. This problem has been noted by Commoner (1990: 16): "If the technosphere is ignored, the environmental crisis can be defined in purely ecological terms. Human beings are then seen as a peculiar species, unique among living things, that is doomed to destroy its own habitat. Thus simplified, the issue attracts simplistic solu-tions: reduce the number of people; limit their share of nature's resources; protect all other species from the human marauder by endowing them with 'rights.'"

2. "Language use" stands for all symbolic communication, including speech, writ-ing, and gestures.

3. This point has been developed by Madison (1981: 108–144), Foucault (1972: 118–125, 215–237), and Saussure (1966: 65–78, 114–120).

4. In this book I adopt Habermas's position on closure as the result of fulfillment of validity claims. There is a major controversy over his notion of universal prag-matics. In opposition to this position, rhetorical thinkers seeks to avoid a develop-ing a foundationalist position on language use (Brown 1994). Neither position has attained a preeminent position in social thought. For an excellent debate on this subject, see Hoy and McCarthy 1994.

5. In one of the most common misunderstandings of this scheme, validity claims are labeled transcendental constructions. Habermas is quite specific in refuting this charge. He maintains that his theory of communicative action is a reconstructive science "for which plausible confirmation must be sought" (Habermas 1990: 116–117). In addition, his application of this theory to social order is one attempt to demonstrate the utility of this framework. The second volume of *The Theory of Communicative Action* (Habermas 1987a) serves to demonstrate this perspective's utility and to contribute to its truth claim (Habermas 1984: 139–140).

6. Discourses create contingent networks of communicative action that define a specific pattern of interaction that forms a community. Network analysis of these communities (Emirbayer and Goodwin 1994; Dietz and Rycroft 1987: 77–101) can provide an empirical test of the existence of communities based on a common dis-course.

7. This point has been well established by a number of authors (Brown 1977; Brown 1987; Greimas 1987; Ricoeur 1981; Habermas 1987a; Burke 1945, 1954; Gadamer 1984). In addition, for a discussion of the narrative structure of knowl-edge verses rule theory, see Habermas 1987a. An application of this structure of knowledge to evolutionary theory can be found in Landau 1991.

8. Habermas's use of systems theory has been a subject of intense debate (McCarthy 1991; Alexander and Turner 1985; Bohman 1989). Habermas has responded to these critiques (1991: 250–264), and his position has been incorporated into this text. A preferred alternative to systems theory may be evolutionary theory (Eder 1993: 17–41; Strydom 1993; Dietz and Burns 1992). The possible incorporation of the theory of communicative action into evolutionary theory is an active and interesting area of theoretical research. However, this project is well beyond the scope and concerns of this text.

9. Clearly, Habermas's use of 'evolution' is based in a developmentalist notion of social change (Dietz et al. 1989)—i.e., in a teleological model of change according to a pre-given logic. This developmentalist logic has been criticized by Eder (1993). In addition, the notion of social learning developed by Habermas has been criticized as an "ontogenic fallacy"—i.e., a conflation of individual learning processes with collective learning (Strydom 1987, 1993; Eder 1993). Habermas (1991: 262) has noted this critique and sees the need to develop a theory of collective learning; he sees Eder's work as "a notable step in this direction." On the development of an evolutionary alternative to developmentalist models of social change, see Dietz and Burns 1992 and Eder 1993.

10. Mead (1934: 309) describes this type of personality as being able to "contemplate himself as a whole: his ability to take the social attitudes of other individuals and also of the generalized other toward himself, within the given organized society of which he is a member, makes possible his bringing himself, as an objective whole, within his own experiential purview, and thus he can consciously integrate and unify the various aspects of his self, to form a single, consistent, and coherent and organized personality. Moreover, by the same means, he can undertake and effect intelligent reconstructions of that self or personality in terms of its relations to the given social order, whenever the exigencies of adaptation to his social environment demand such reconstructions."

11. For summaries of the epistemological arguments on which the theory of communicative action is based, see Radnitzky 1973 and Bleicher 1982.

12. Even this notion of rationality is extremely limited within state institutions (Dutton et al. 1980).

13. Eckersley (2000: 18) defines the Precautionary Principle in the following manner: "When there are threats of serious or irreversible environmental damage, then lack of certainty should not be taken as a reason for postponing or avoiding taking measures to prevent the damage."

14. This parallels discussions of the development of ecological ethics by Merchant (1992) and by Stern, Dietz, and Kalof (1992).

15. One possible means of representation of nonlinguistic beings in human decision making was argued by Justice William O. Douglas in a 1971 Supreme Court case, *Sierra v. Morton*. Douglas argued for the appointment of guardians that would represent the interests of the natural environment in government proceedings: "The critical question of standing would be simplified and also put neatly in focus if we fashioned a federal rule that allowed environmental issues to be litigated before

federal agencies or federal courts in the name of the inanimate object about to be despoiled, defaced, or invaded by roads and bulldozers and where injury is the subject of public outrage. . . . The sole question is, who has standing to be heard? . . . Permitting a court to appoint a representative of an inanimate object would not be significantly different from customary judicial appointments of guardians ad litem, executors, conservators, receivers, or counsel for indigents. . . . That is why these environmental issues should be tendered by the inanimate object itself. Then there will be assurances that all of the forms of life which it represents will stand before the court—the pileated woodpecker as well as the coyote and bear, the lemmings as well as the trout in the streams." (Douglas 1971)

Chapter 3

1. See chapter 2 above.

2. See Buttel 1986, Buttel 1987, and Dunlap 1993.

3. Beck's analysis is, perhaps, one of the most comprehensive.

4. For summaries of these arguments, see Oelschlaeger 1992a,b and Elliot and Gare 1983.

5. On patriarchy and the creation of environmental problems, see Diamond and Orenstein 1990. For examples of other analyses pointing to various cultural causes of environmental degradation, see Jaeger 1994; Merchant 1980; White 1967; Teymur 1982; Meeker 1972; Evernden 1992b.

6. See chapter 2 above.

7. See Graham and Sadowitz 1994; Dietz 1987, 1988; Fischer 1990; Forester 1985; Stanley 1983; Strassman 1978.

8. One of the more interesting and theoretically informed examples of this type of institutional experimentation has been the development of citizen participation in hazardous-waste siting decisions in Switzerland (Webler 1992; Webler, Kastenholz, and Renn 1994).

9. The debate on environmental ethics in the United States is summarized in Nash 1989b. For summaries of key readings on this subject, see Elliot and Gare 1983 and Oelschlaeger 1992a.

10. See chapter 2 above.

11. The links between social learning, democracy, and the resolution of environmental degradation have been noted by a number of scholars. For example, Brown (1989: 148–149) notes that the resolution of our social structures with needs developed in the lifeworld lies in the restoration of a democratic political order (ibid.: 144). O'Riordan (1980: 428) also identifies public participation as one necessary component of this strategy to ensure the accountability of government. Schnaiberg (1980: 424) extends this argument by focusing on the need to control the market to resolve environmental problems. Jaeger (1994: 273–274) combines these concerns over the administrative state and the market. He dismisses the technocratic

managed society as a viable option to deal with environmental problems, and stresses the need for human agency in the development of cultural innovations to deal with environmental problems. Paehlke (1988: 304–309) argues for more public participation in bureaucratic decision making

12. Among the key theorists are Habermas (1987), Cohen (1982), Keane (1988), Ely (1992), Calhoun (1993), and Torgerson (1999).

13. There is an extensive literature on the idea of civil society. For a summary of the leading positions on this topic, see Livesay 1994. This discussion follows the notion of civil society developed by Habermas (1987) and expanded on by Cohen and Arato (1992) and by Ely (1992).

Chapter 4

1. For a summary of this perspective, see Stewart et al. 1989. For recent analyses of this approach to social movements, see Benford and Hunt 1992 and Brulle 1994.

2. Dramatism defines a approach to the study of social action which maintains that "the most direct route to the study of human relations and human motives is via a methodological inquiry into cycles or clusters of terms and their functions" (Burke 1968: 445). For descriptions of the dramatistic analysis of social movements, see Burke 1945, 1955, 1989; Griffin 1966; Stewart et al. 1989.

3. For descriptions of this process, see Bruner 1987; Brown 1977, 1987; Habermas 1987, 1992; Burke 1945, 1954.

4. There is a distinct but uncalculated convergence between the work of Benford and Habermas on the acceptance of a discursive frame (Snow and Benford 1988; Habermas 1984). As I noted in chapter 2, to establish this resonance and acceptance by an individual, a discourse must fulfill the validity claims of truth, normative correctness, and sincerity. Benford separately arrived at a similar scheme for the criteria of acceptance of a discursive frame. This scheme is based on its empirical credibility, narrative fidelity, and experiential commensurability (Snow and Benford 1988: 207–211; 1992: 140–141).

5. The literature utilizing framing analysis is expanding rapidly and cannot be adequately summarized here. For the development and use of this perspective, see Gamson 1988, 1991, 1992; Klandermans 1992; Snow and Benford 1988, 1992; Snow et al. 1986. For an analysis of worldviews and tactics for social change, see Dietz et al. 1989. For a recent application of this type of analysis to social movements in Germany, see Gerhards and Rucht 1992.

6. For fuller descriptions of this model, see Zald and Garner 1987 and Eyerman and Jamison 1991.

7. For examples of this type of analysis, see Milbrath 1984, 1989; Schnaiberg and Gould 1994.

8. For a further analysis using this line of argument to examine environmental sociology, see Routley 1983.

9. Douglas and Wildavksy (1982) divide environmental attitudes into four unchanging categories—fatalism, egalitarianism, hierarchy, and individualism— that "coexist in any social entity" (Dake and Thompson 1993: 422). Similar analyses read the U.S. environmental movement as a quest for beauty, health, and permanence (Hays 1985, 1987), as a progressive evolution of "ethics from the natural rights of a limited group of humans to the rights of nature" (Nash 1989b: 4), or as divided into problem solving, survivalism, sustainability, and "green radicalism" (Dryzek 1997).

10. Some examples of this approach include examinations of radical movement organizations (Scarce 1990; Manes 1990; Luke 1994), of the environmental justice movement (Szasz 1994; Schwab 1994), of "not in my back yard" organizations (Freudenberg 1984; Freudenberg and Steinsapir 1992), and of the People of Color Environmental Movement (Bullard 1994a; Gedicks 1993). Studies of aggregates of environmental organizations included Mitchell 1989, Mitchell 1992, Ingram and Mann 1989, and Tober 1989.

11. Several scholars have examined the development of the "new" social movements. The key authors are Adorno, Horkheimer, Marcuse, Cohen, Arato, Offe, Dalton, Kuechler, Burklin, Strydom, Habermas, Offe, and Eder. For an examination of the roles of Marcuse, Horkheimer, and Adorno in the development of the idea of "new" social movements, see Scheffer 1991: 18.

12. In this book I do not attempt to summarize the development of the concept of biodiversity. For excellent discussions of the social construction of this term, see Takacs 1996 and Grumbine 1997.

13. For the literature to support this statement, see Benford and Hunt 1992; Zald and Garner 1987; Dietz, Stern, and Rycroft 1989; Knoke 1990; Gamson 1991; Snow and Benford 1988.

14. The initial source of the discourse categories used here is Oelschlaeger 1991. Other authors have developed similar listings. For example, there is virtual accord in the environmental-history community (Nash 1966; Worster 1973; Hays 1959) that the two early environmental discourses in the U.S. were Conservationism and Preservationism. A discourse similar to what is labeled Reform Environmentalism is mentioned by several authors, who use various terms such as *biocentrism, envirocentrism,* and *the new environmentalism* (O'Brien 1983; McCormick 1989; Scheffer 1991). In addition, Deep Ecology and Ecofeminism have been identified as significant perspectives in American environmentalism (Merchant 1992; Roussopoulos 1993; Zimmerman 1994). This listing of environmental discourses does not exhaust the possibility of including other discourses, including alternative technology (Dickson 1975; Willoughby 1990; Friedman 1992). But it is not included as a discourse that represents a significant community in the U.S. environmental movement. There is no barrier in the theoretical framework that would preclude inclusion of alternative technology. But this discourse does not seem to animate a significant level of collective behavior. Accordingly, only the nine environmental discourses listed above are examined in detail because they provide an identity to the overwhelming majority of U.S. environmental organizations.

Chapter 5

1. Source: *Statistical Abstract of the U.S.* 1997, table 14.

Chapter 6

1. See also Burch 1971: 119.

2. Two early examples of this debate are Chapman 1913 and Lamb 1913.

3. A portion of this speech is included in the 1993 film *Rachel Carson*.

4. The CDFE (headquartered in Bellevue, Washington, and led by Ron Arnold) emerged as a central figure in the wise-use movement through its sponsorship of the Reno meeting. It continues to serve as a coordination and publication center for the movement.

5. See the appendix for descriptions of how this analysis was performed.

6. Averages of years are arithmetic means.

7. *Parliamentarian*: working through existing government bodies.

Chapter 7

1. Source of quotation: Grinnel and Sheldon 1895: 224.

2. Source of quotation: Trefethen 1964: 16.

3. For a more detailed analysis of these groups, see Tober 1989.

4. Marsh was an avid opponent of capitalism's immoral abuse of the natural environment. For an extended discussion of the role of private enterprise in the creation of ecological destruction, see Marsh 1864: 51–52.

5. Fernow was also chairman of the AFA's executive committee (Dana and Fairfax 1980: 43).

6. In June of 1908, the president issued an executive order creating a National Conservation Commission to carry out the recommendations of the conference (Van Hise 1921: 182). The first meeting of this group occurred in August of 1909 (Proceedings of the First National Conservation Congress 1909).

7. This is commonly called the Brundtland Report (after Gro Harlem Brundtland, who headed the commission).

8. This is how the name was styled at the time.

9. Included are the following organizations (Bent 1916: 5–18; Jones 1965: 4–5):

1873 White Mountain Club
1875 Rocky Mountain Club
1892 Sierra Club
1893 Mazamas
1895 American Scenic and Historical Preservation Society

1895 American Park and Outdoor Art Society
1900 Association for the Protection of the Adirondacks
1900 Society for the Preservation of Historical and Scenic Spots
1901 American Alpine Club
1901 Mt. Whitney Club
1906 Seattle Mountaineers
1910 Green Mountain Club

10. An excellent example of this type of publication is *Celebrating the American Earth—A portfolio by Ansel Adams*, issued by the Wilderness Society.

11. Based on Nature Conservancy Information Tax Return (IRS 990) for 1994.

Chapter 8

1. See chapter 3 of the present volume.

Chapter 9

1. For discussions of this discourse, see Devall and Sessions 1985; Oelschlaeger 1991; Short 1991.

2. David Brower is a member (and an elder statesman) of this coalition.

3. See *Earth First! Journal* 14 (1994), no. 5: 37–38.

4. The Cobb symbol (resembling the Greek capital theta) is a symbolic representation of the ecology movements. It was developed by Ron Cobb in a cartoon that appeared in the *LA Free Press* in the fall of 1969. For a history of the development of this symbol, see *Environmental Action* 21, no. 4 (1990).

5. For a detailed discussion and qualitative analysis of the use of this term, see Capek 1996.

6. See Gibbs 1982; Cable and Cable 1995: 75–84.

7. Among these groups were the Indigenous Environmental Network, the Southern Organizing Committee, and the Southwest Network for Environmental and Economic Justice (Ferris and Hahn-Baker 1995: 69). See also Bullard 1994a.

8. Some of the main network organizations are the Asian Pacific Environmental Network, the Farmworker Network of Economic and Environmental Justice, the Indigenous Environmental Network, the Northeast Environmental Justice Network, the Southern Organizing Committee for Economic and Social Justice, the Southwest Network for Environmental and Economic Justice, and the Citizens' Clearinghouse for Hazardous Waste.

9. See Ortner 1974 and Ruther 1975.

10. See Oelschlaeger 1991: 309.

11. These statements are reprinted in Gottlieb 1996.

Chapter 10

1. On the selection of these 87 organizations, see the appendix.

2. See also chapter 4 of the present volume.

3. An analysis of the practices of the Nature Conservancy might result in a change in its categorization. The annual meeting was held in the Ryetown Hilton, in a suburb of New York City, on a Friday afternoon. Member input to the voting was either by attending the meeting personally, or sending in an "Official Ballot" which reads "I cast my vote for the slate of officers and governors proposed by the nominating committee." While suggestive of an oligarchic organizational practice, further analysis is needed to ensure a correct recategorization of this organization's governmental form. This re-categorization would magnify the extent of oligarchic practices within environmental movement organization.

4. See chapter 4 above.

5. See chapter 4 above.

Chapter 11

1. Source: Habermas 1994: 97.

2. See chapter 3 above.

3. Harre et al. (1999) develop a similar definition under the notion of "greenspeak."

4. See chapter 2 above.

Bibliography

Abbey, Edward. 1975. *The Monkey Wrench Gang*. Avon.

Abernethy, Virginia. 1998. Defining the new American community: A slide to tribalism? In *The Coming Age of Scarcity*, ed. M. Dobkowski and I. Wallimann. Syracuse University Press.

Abramovitz, Janet N. 1998. Sustaining the world's forests. In *State of the World*, ed. L. Brown. Norton.

Adams, Ansel. No date. *Celebrating the American Earth*. Wilderness Society.

Adorno, Theodor, and Horkheimer, Max. 1944. *Dialectic of Enlightenment*. Continuum.

Adler, Jonathan. 1995. *Environmentalism at the Crossroads: Green Activism in America*. Capital Research Center.

Alexander, Jeffrey, and Turner, Jonathan, eds. 1985. *Neofunctionalism*. Sage.

Allen, Thomas B. 1987. *Guardian of the Wild: The Story of the National Wildlife Federation, 1936–1986*. Indiana University Press.

Alliance for the Wild Rockies. 1995. Northern Rockies Ecosystem Protection Act.

American Association for Promoting Hygiene and Public Baths. 1916. Address by Dr. Simon Baruch. In Proceedings of Fifth Annual Meeting, Baltimore.

Amy, Douglas J. 1987. *The Politics of Environmental Mediation*. Columbia University Press.

Anderson, Nancy K., and Ferber, Linda S. 1990. *Albert Bierstadt: Art and Enterprise*. Hudson Hills.

Andrews, James R. 1980. History and theory in the study of the rhetoric of social movements. *Central States Speech Journal* 31, no. 4: 274–281.

Andrews, Richard N. L. 1999. *Managing the Environment, Managing Ourselves: A History of American Environmental Policy*. Yale University Press.

Anonymous. 1783. *The Sportsman's Companion, or an Essay on Shooting*. Isaak Nealf.

Apel, K. O. 1980. *Toward a Transformation of Philosophy*. Routledge and Kegan Paul.

Aronowitz, Stanley. 1988. *Science as Power: Discourse and Ideology in Modern Society*. University of Minnesota Press.

AUA (American Unitarian Association). 1910. Conservation of National Resources. Social Service Series Bulletin No. 28.

Bailey, L. H. 1915. *The Holy Earth*. Scribner.

Baldwin, Donald N. 1972. *The Quiet Revolution: Grass Roots of Today's Wilderness Preservation Movement*. Pruett.

Barnouw, Erik. 1975. *Tube of Plenty: The Evolution of American Television*. Oxford University Press

Bartlett, John. 1968. *Familiar Quotations*. Little, Brown.

Bean, Michael J. 1978. Federal wildlife law. In *Wildlife and America*, ed. H. Brokaw. Council on Environmental Quality.

Beck, Ulrich. 1992a. From industrial to risk society. *Theory, Culture and Society* 9, no. 1: 97–123.

Beck, Ulrich. 1992b. *Risk Society: Towards a New Modernity*. Sage.

Beck, Ulrich. 1995. *Ecological Enlightenment: Essays on the Politics of the Risk Society*. Humanities.

Beck, Ulrich. 1996. World risk society as cosmopolitan society? Ecological questions in a framework of manufactured uncertainties. *Theory, Culture and Society* 13, no. 4: 1–32.

Bellah, Robert N., Richard Madsen, W. M. Sullivan, A. Swindler, and S. M. Tipton. 1985. *Habits of the Heart*. Harper and Row.

Bendavid, Naftali. 1996. Environmental inaction. *Legal Times*, December 2.

Benford, Robert D., and Scott A. Hunt. 1992. Dramaturgy and social movements: The social construction and communication of power. *Sociological Inquiry* 62, no 1: 36–55.

Benhabib, Seyla, and Fred Dallmayr. 1990. *The Communicative Ethics Controversy*. MIT Press.

Bennett, H. H., and W. R. Chapline. 1929. Soil Erosion: A National Menace. Circular 33, U.S. Department of Agriculture.

Bent, Allen H. 1916. The mountaineering clubs of America. *Appalachia*, December: 5–18.

Bentham, Jeremy. 1789. An introduction to the principles of morals and legislation. In *The Utilitarians* (Anchor Books, 1973).

Benton, T. 1994. Biology and social theory in the environmental debate. In Benton and Redclift 1994.

Benton, T., and M. Redclift, eds. 1994. *Social Theory and the Global Environment*. Routledge.

Bernstein, Richard J., ed. 1985. *Habermas and Modernity*. MIT Press.

Bittner, E. 1965. The concept of organization. In *Studies in Ethnomethodology*, ed. R. Turner (Penguin, 1974).

Blalock, Hubert M. 1979. *Social Statistics*, revised second edition. McGraw-Hill.

Bleicher, Josef. 1980. *Contemporary Hermeneutics*. Routledge and Kegan Paul.

Bleicher, Josef. 1982. *The Hermeneutic Imagination*. Routledge and Kegan Paul.

Bohman, James. 1989. 'System' and 'lifeworld': Habermas and the problem of holism. *Philosophy and Social Criticism* 15, no. 4: 381–401.

Bookchin, Murray. 1962. *Our Synthetic Environment*. Knopf.

Bookchin, Murray, Dave Foreman, Steven Chase, and David Levine. 1991. *Defending the Earth*. South End.

Borecelli, Peter. 1988. *Crossroads*. Island.

Borowsky, Larry. 1990. Dusting off the Cobb Symbol. *Environmental Action* 21, no. 4: 7–8.

Bottomore, T. B. 1970. Elites and society. In *Power in Societies*, ed. M. Olsen. Macmillan.

Bouchier, D. 1986. The sociologist as anti-partisan: A dilemma of knowledge and academic power. *Research in Social Movements, Conflict, and Change* 9: 1–23.

Bowers, William L. 1974. *The Country Life Movement in America 1900–1920*. Kennikat.

Braudel, F. 1980. *On History*. University of Chicago Press.

Brick, P., and R. Cawley. 1996. Knowing the wolf, tending the garden. In *A Wolf in the Garden*, ed. Brick and Cawley. Rowman and Littlefield.

Brinton, Alan. 1985. On viewing knowledge as rhetorical. *Central States Speech Journal* 36: 270–281.

Bronner, Stephen Eric. 1994. *Of Critical Theory and Its Theorists*. Blackwell.

Brown, Lester. 1991. The Battle for the Planet: A Status Report. In *Environment in Peril*, ed. A. Wolbarst. Smithsonian.

Brown, Richard H. 1977. *A Poetic for Sociology*. Cambridge University Press.

Brown, Richard H. 1978. Bureaucracy as praxis: Toward a political phenomenology of formal organizations. *Administrative Science Quarterly* 23, September: 365–382.

Brown, Richard H. 1983. Theories of rhetoric and the rhetoric of theory: Toward a political phenomenology of sociological truth. *Social Research* 50, no. 1: 126–157.

Brown, Richard H. 1987. *Society as Text*. University of Chicago Press.

Brown, Richard H. 1989. *Social Science as Civic Discourse*. University of Chicago Press.

Brown, Richard H. 1990. Rhetoric, textuality, and the postmodern turn in sociological theory. *Sociological Theory* 8, no. 2: 188–197.

Brown, Richard H. 1994. Reconstructing social theory after the postmodern critique. In *After Postmodernism*, ed. H. Simons and M. Billig. Sage.

Brown, Richard H. 1998. *Toward a Democratic Science: Scientific Narration and Civic Communication*. Yale University Press.

Brubaker, R. 1985. Rethinking classical theory. Theory *and Society* 14, no. 6: 745–776.

Brulle, Robert J. 1992. Jürgen Habermas: An exegesis for human ecologists. *Human Ecology Bulletin* no. 8: 29–40.

Brulle, Robert J. 1994. Power, discourse, and social problems: Social problems from a rhetorical perspective. *Current Perspectives in Social Problems* 5: 95–121.

Brulle, Robert J. 1995a. Agency, Democracy and the Environment. Ph.D. dissertation, Department of Sociology, George Washington University.

Brulle, Robert J. 1995b. Environmentalism and human emancipation. In *Social Movements*, ed. S. Lyman. Macmillan.

Brulle, Robert J. 1996. Environmental discourse and social movement organizations: A historical and rhetorical perspective on the development of U.S. environmental organizations. *Sociological Inquiry* 65, no. 4: 58–83.

Brulle, Robert J., and Richard H. Brown. 1997. Democratic science in practice: The experience of the environmental justice movement. In R. Brown, *Science as Narration*. Yale University Press.

Brulle, Robert J., and Thomas Dietz. 1992. Greening Rhetoric: Beyond Anthropocentrism in Communication Studies. Presented at Conference on Academic Knowledge and Political Power, University of Maryland.

Brulle, Robert J., and Thomas Dietz. 1993. A Rhetoric for Nature: Beyond the Anthropocentric/Naturalism Dichotomy. Presented at International Conference on The Social Functions of Nature, Chantilly, France.

Brummett, B. 1976. Some implications of process or intersubjectivity: Postmodern rhetoric. *Philosophy and Rhetoric* 9, no. 1: 21–51.

Bruner, J. 1987. Life as narrative. *Social Research* 54, no. 1: 11–32.

Bruner, Michael, and Max Oelschlaeger. 1994. Rhetoric, environmentalism, and environmental ethics. *Environmental Ethics* 16, winter: 377–396.

Bryant, Bunyan, ed. 1995. *Environmental Justice*. Island.

Bryant, Bunyan, and Paul Mohai, eds. 1992. *Race and the Incidence of Environmental Hazards*. Westview.

Buechler, Steven M. 1993. Beyond resource mobilization? Emerging trends social movement theory. *Sociological Quarterly* 34, no. 2: 217–235.

Bullard, Robert D. 1990. *Dumping in Dixie: Race, Class and Environmental Quality*. Westview.

Bullard, Robert D. 1993. Anatomy of environmental racism and the environmental justice movement. In *Confronting Environmental Racism*, ed. R. Bullard. South End.

Bullard, Robert D. 1994a. People of Color Environmental Groups, 1994–1995 Directory. Environmental Justice Resource Center, Atlanta.

Bullard, Robert D. 1994b. *Unequal Protection; Environmental Justice and Communities of Color*, Sierra Club.

Bullard, Robert D., and Beverly H. Wright. 1992. The quest for environmental equity: Mobilizing the African-American community for social change. In *The U.S. Environmental Movement, 1970–1990*, ed. R. Dunlap and A. Mertig. Taylor and Francis.

Burch, William R., Jr. 1971. *Daydreams and Nightmares: A Sociological Essay on the American Environment*. Harper and Row.

Burch, William R., Jr. 1976 The peregrine falcon and the urban poor: Some sociological interrelations. In *Human Ecology*, ed. P. Richerson and J. McEvoy III. Duxbury.

Burke, Kenneth. 1945, *A Grammar of Motives*. University of California Press.

Burke, Kenneth. 1954, *Permanence and Change*. University of California Press.

Burke, Kenneth. 1989. *On Symbols and Society*. University of Chicago Press.

Burns, Tom R., and Thomas Dietz. 1992. Cultural evolution: Social rule systems, selection, and human agency. *International Sociology* 7: 250–283.

Buttel, F. 1986. Sociology and the environment: The winding road toward human ecology. *International Social Science Journal*, no. 109: 335–356.

Buttel, F. 1987. New directions in environmental sociology. *Annual Review of Sociology* 13: 465–488.

Buttel, F., and Taylor, P. 1994. Environmental Sociology and Global Environmental Change: A Critical Assessment. In *Social Theory and the Global Environment*, ed. T. Benton and M. Redclift. Routledge.

Cable, Sherry, and Charles Cable. 1995. *Environmental Problems: Grassroots Solutions*. St. Martin's Press.

Calhoun, Craig. 1987. Populist Politics, Communications Media and Large Scale Social Integration. Working Papers and Proceedings of the Center for Psychosocial Studies, no. 16.

Calhoun, Craig. 1993. Nationalism and civil society: Democracy, diversity and self-determination. *International Sociology* 8, no. 4: 387–411.

Calhoun, Craig. 1994. Postmodernism as pseudohistory: Continuities in the complexities of social action. In *Agency and Structure*, ed. P. Sztompka. Gordon and Breach.

Calhoun, Craig. 1995. *Critical Social Theory: Culture, History, and the Challenge of Difference*. Blackwell.

Callahan, D., and Myer, R. 1992. The Wise Use Movement. Working Paper, W. Alton Jones Foundation, Charlottesville, Virginia.

Camhis, Marios. 1979. *Planning Theory and Philosophy*. Tavistock.

Canan, Penelope, and George W. Pring. 1988. Strategic lawsuits against public participation. *Social Problems* 35: 506–519.

Capek, Stella M. 1996. The "environmental justice" frame: A conceptual discussion and an application. *Human Ecology Review* 2 , no. 2: 114–131.

Capra, Fritjof, and Charlene Spretnak. 1984. *Green Politics: The Global Promise*. Dutton.

Carleheden, M., and R. Gabriels. 1996. An interview with Jürgen Habermas. *Theory, Culture and Society* 13 (3): 1–17.

Carney, Eliza Newlin, and W. Hohn Moore. 1992. From the K Street corridor. *National Journal*, January 4.

Carson, Rachel. 1951. *The Sea around Us*. Oxford University Press.

Carson, Rachel. 1962. *Silent Spring*. Riverside.

Caterino, Brian. 1994. Communicative ethics and the claims of environmental justice. Presented at Annual Conference of the American Political Science Association, Washington.

Cathcart, Robert S. 1972. New approaches to the study of movements: Defining movements rhetorically. *Western Speech*, spring: 82–88.

Cathcart, Robert S. 1978. Movements: Confrontation as rhetorical form. *Southern Speech Communication Journal* 43: 233–247.

Cathcart, Robert S. 1980. Defining social movements by their rhetorical form. *Central States Speech Journal* 31, no. 4: 267–273.

Catlin, George. 1833 (1844). *Letters and Notes on the Manners, Customs, and Conditions of the North American Indians*. Dover.

Catton, W. R. 1980. *Overshoot: The Ecological Basis of Revolutionary Change*. University of Illinois Press.

Catton, W. R., and R. E. Dunlap. 1978. Environmental sociology: A new paradigm? *American Sociologist* 13: 41–49.

Catton, W. R., and R. E. Dunlap. 1980. A new ecological paradigm for post exuberant sociology. *American Behavioral Scientist* 24: 14–57.

Caulfield, Henry P. 1989. The conservation and environmental movements: An historical analysis. In *Environmental Politics and Policy*, ed. J. Lester. Duke University Press.

Cawley, R. McGreggor. 1993. *Federal Land, Western Anger: The Sagebrush Rebellion and Environmental Politics*. University Press of Kansas.

CCHW (Citizens' Clearinghouse for Hazardous Waste). 1986. First National Grassroots Convention. *Everybody's Back Yard* 4, no. 3: 1–2.

Cerrell Associates Inc. 1984. *Political Difficulties Facing Waste-to-Energy Conversion Plant Siting*.

Chadsey, Mildred. 1915. Municipal housekeeping. *Journal of Home Economics* 7, February: 53–59.

Chadwick, Edwin. 1842. Report from the Poor Law Commissioners on an Inquiry into the Sanitary Conditions of the Labouring Population of Great Britain.

Chapman, Herman H. 1913. Shall the national forests be turned over to the states? *American Forestry*, January: 79–87.

Chivian, Eric, Michael McCally, Howard Hu, and Andrew Haines, eds. 1993. *Critical Condition: Human Health and the Environment*. MIT Press.

Clarke, Robert. 1973. *Ellen Swallow: The Woman Who Founded Ecology*. Follett.

Clearinghouse on Environmental Advocacy and Research. 1998. *A Clear View*. Volume 5, no. 13.

Clepper, Henry. 1966. *Origins of American Conservation*. Ronald.

Clepper, Henry, ed. 1971. *Leaders of American Conservation*. Ronald.

Cockburn, A., and J. St. Clair. 1994. Death and life for America's Greens. *The Nation*, December 19, 1994.

Cohen, Jean L. 1982. *Class and Civil Society: The Limits of Marxian Critical Theory*. University of Massachusetts Press.

Cohen, Jean L. 1985. Strategy or identity: New theoretical paradigms and contemporary social movements. *Social Research* 52, no. 4: 663–716.

Cohen, Jean L., and Arato, A. 1992. Politics and the reconstruction of the concept of civil society. In *Cultural-Political Interventions in the Unfinished Project of Enlightenment*, ed. A. Honneth et al. MIT Press.

Colwell, Mary A. C. 1993. *Private Foundations and Public Policy: The Political Role of Philanthropy*. Garland.

Commoner, Barry. 1970. Fundamental causes of the environmental crisis. In *American Environmentalism*, ed. R. Nash. McGraw-Hill.

Commoner, Barry. 1971. *The Closing Circle; Nature, Man, and Technology*. Knopf.

Commoner, Barry. 1990 *Making Peace With the Planet*. Pantheon.

Commoner, Barry. 1991. The failure of the environmental effort. In *Environment in Peril*, ed. A. Wolbarst. Smithsonian.

Cook, John R. 1992. Beyond the Green: Redefining and Diversifying the Environmental Movement. Environmental Careers Organization, Boston.

Cooper, Marilyn M. 1996. Environmental rhetoric in the age of hegemonic politics: Earth First! and the Nature Conservancy. In *Green Culture*, ed. C. Herndl and S. Brown. University of Wisconsin Press.

Cooper, Martha. 1988. Rhetorical criticism and Foucault's philosophy of discursive events. *Central States Speech Journal* 39, no. 1: 1–17.

Cosgrove, Denis. 1994. Contested global visions: One-world, whole-Earth, and the Apollo space photographs. *Annals of the Association of American Geographers* 84, no. 2: 270–294.

Cotgrove, Stephen. 1982. *Catastrophe or Cornucopia: The Environment, Politics and the Future*. Wiley.

Cotgrove, Stephen, and Duff, Andrew. 1980. Environmentalism, middle-class radicalism, and politics. *Sociological Review* 28, no. 2: 333–352.

CEQ (Council on Environmental Quality). 1971. *Environmental Quality: The Second Annual Report of the Council on Environmental Quality.* Government Printing Office.

Council of Hygiene and Public Health, Citizens' Association of New York. 1866. *Report upon the Sanitary condition of the City.* Appleton.

Cox, J. Robert. 1981. Argument and the definition of the situation. *Central States Speech Journal* 32, fall: 197–205.

Crandall, Robert W. 1987. Learning the lessons. *Wilson Quarterly*, autumn: 69–80.

Crevecoeur, J. Hector St. John de. 1782. Letters from an American Farmer. Printed for T. Davies.

Cronon, William. 1983. *Changes in the Land: Indians, Colonists, and the Ecology of New England.* Hill and Wang.

Crowfoot, James E. 1992. Conservation leadership in academia. In *Voices from the Environmental Movement*, ed. D. Snow. Island.

Crowfoot, James E., and Bunyan I. Bryant Jr. 1980. Environmental Advocacy: An Action Strategy For Dealing With Environmental Problems. *Journal of Environmental Education* 11, no. 3: 36–41.

Culler, J. 1985. Communicative competence and normative force. *New German Critique* 35, spring-summer: 133–144.

Cuthbert, Thomas N. 1889, The New York Association for the Protection of Game. *Forest and Stream*, no. 33, December 26.

Cutter, Susan L. 1994. Environmental issues: Green rage, social change and the new environmentalism. *Progress in Human Geography* 18, no. 2: 217–226.

Dake, Karl, and Michael Thompson. 1993. The meanings of sustainable development: Household strategies for managing needs and resources. In *Proceeding of the Society for Human Ecology Conference, 1993.* Society for Human Ecology.

Dalton, Russell J., and Manfred Kuechler. 1990. *Challenging the Political Order: New Social and Political Movements in Western Democracies.* Oxford University Press.

Dalton, Russell J., Manfred Kuechler, and Wilhelm Bürklin. 1990. The challenge of new movements. In *Challenging the Political Order*, ed. R. Dalton and M. Kuechler. Oxford University Press.

D'Amico, Robert. 1991. The myth of the totally administered society. *Telos* 88, summer: 80–96.

Dana, S. T., and S. K. Fairfax. 1980. *Forest and Range Policy: Its Development in the United States.* McGraw-Hill.

Daniel, Joseph E., et. al. 1991. *1992 Earth Journal—Environmental Almanac and Resource Directory.* Buzzworm.

d'Eaubonne, F. 1974. The time for ecofeminism. Reprinted in *Ecology*, ed. C. Merchant and R. Gottlieb (Humanities, 1994).

Defenders of Wildlife and Sierra Club Legal Defense Fund. 1997. Endangered Natural Heritage Act: Situations and Solutions. Defenders of Wildlife.

Devall, Bill. 1970. The Governing of a Voluntary Organization: Oligarchy and Democracy in the Sierra Club. Ph.D. dissertation, University of Oregon.

Devall, Bill. 1992. Deep ecology and radical environmentalism. In *The U.S. Environmental Movement, 1970–1990*, ed. R. Dunlap and A. Mertig. Taylor and Francis.

Devall, Bill, and George Sessions. 1985. *Deep Ecology: Living As If Nature Mattered*. Peregrine Smith Books.

DeVoto, Bernard. 1948. Sacred Cows and Public Lands. *Harpers*, July: 44–55.

DeWitt, Calvin B., ed. 1991. *The Environment and the Christian: What Does the New Testament Say about the Environment?* Baker Book House.

Diamond, Irene, and Gloria Feman Orenstein, eds. 1990. *Reweaving the World: The Emergence of Ecofeminism*. Sierra Club.

Diani, M., and M. Eyerman. 1992. *Studying Collective Action*. Sage.

Dickson, David. 1975. *The Politics of Alternative Technology*. Universe Books.

Dietz, Thomas. 1987. Theory and method in social impact assessment. *Sociological Inquiry* 57(1): 54–69.

Dietz, Thomas. 1988. Social impact assessment as applied human ecology: Integrating theory and method. In *Human Ecology*, ed. R. Borden et al. Society for Human Ecology.

Dietz, Thomas, and T. R. Burns. 1992. Human agency and the evolutionary dynamics of culture. *Acta Sociologica* 35: 187–200.

Dietz, Thomas, and Eugene A. Rosa. 1994. Rethinking the environmental impacts of population, affluence and technology. *Human Ecology Review* 1, summer-autumn: 277–300.

Dietz, Thomas, and Eugene A. Rosa. 1997. Environmental impacts of population and consumption. In *Environmentally Significant Consumption*, ed. P. Stern et al. National Academy Press.

Dietz, Thomas, and Robert W. Rycroft. 1987. *The Risk Professionals*. Russell Sage Foundation.

Dietz, Thomas, T. R. Burns, and F. H. Buttel. 1989. Evolutionary Theory in Sociology: An Examination of Current Thinking. Unpublished manuscript, George Mason University.

Dietz, Thomas, T. R. Burns, and F. H. Buttel. 1989. Evolutionary theory in sociology: An examination of current thinking. In *Handbook of Environmental Sociology*, ed. R. Dunlap and W. Michelson. Greenwood.

Dietz, Thomas, Stern, P., and Rycroft, R. 1989. Definitions of conflict and the legitimation of resources: The case of environmental risk. *Sociological Forum* 4: 47–70.

Dobkowski, Michael N., and Isidor Wallimann, eds. 1998. *The Coming Age of Scarcity: Preventing Mass Death and Genocide in the Twenty-First Century.* Syracuse University Press.

Dobson, Andrew. 1993. Critical theory and Green politics. In *The Politics of Nature*, ed. A. Dobson and P. Lucardie. Routledge.

Dobson, Andrew. 1996. Democratising Green theory: Preconditions and principles. In *Democracy and Green Political Thought*, ed. B. Doherty and M. de Geus. Routledge.

Donati, Paolo R. 1992. Political discourse analysis. In *Studying Collective Action*, ed. M. Diani and M. Eyerman. Sage.

Doughty, T. 1830. Characteristics of a true sportsman. Reprinted in *Classics of the American Shooting Field*, ed. J. Phillips and L. Hill (Houghton Mifflin, 1930).

Douglas, Mary, and Aaron Wildavsky. 1982. *Risk and Culture.* University of California Press.

Douglas, W. O. 1971. Separate Opinion, *Sierra Club v. Morton. United States Supreme Court Reports, October Term, 1971.* Lawyers Co-operative.

Dowie, Mark. 1995. *Losing Ground.* MIT Press.

Dreyfus, Hubert L, and Paul Rabinow. 1982 *Michel Foucault: Beyond Structuralism and Hermeneutics.* Northwestern University Press.

Dryzek, John S. 1987. *Rational Ecology: Environment and Political Economy.* Blackwell.

Dryzek, John S. 1990. Green reason: Communicative ethics for the biosphere. *Environmental Ethics* 12, no. 3: 195–210.

Dryzek, John S. 1995. Critical theory as a research program. In *The Cambridge Companion to Habermas*, ed. S. White. Cambridge University Press.

Dryzek, John S. 1997. *The Politics of the Earth: Environmental Discourses.* Oxford University Press.

Dryzek, John S., and Douglas Torgerson. 1993. Democracy and the policy sciences: A progress report. *Policy Sciences* 26, no. 3: 127–137.

Dubiel, Helmut. 1985. *Theory and Politics: Studies in the Development of Critical Theory.* MIT Press.

Duncan, Otis Dudley. 1961. From social system to ecosystem. *Sociological Inquiry* 31: 140–149.

Dunlap, Riley E. 1980. Paradigmatic change in social science: From human exceptionalism to an ecological paradigm. *American Behavioral Scientist* 24: 5–14.

Dunlap, Riley E. 1991. Public opinion in the 1980s: Clear consensus, ambiguous commitment. *Environment* 33, no. 8: 10–15.

Dunlap, Riley E. 1992. From environmental to ecological problems. In *Social Problems*, ed. C. Calhoun and G. Ritzer. McGraw-Hill.

Dunlap, Riley E., and William R. Catton Jr. 1994. Toward an ecological sociology: The development, current status, and probable future of environmental sociology. In *Ecology, Society, and the Quality of Social Life*, ed. W. D'Antonio et al. Transaction.

Dunlap, Riley E., and Angela G. Mertig, eds. 1992. *The U.S. Environmental Movement, 1970–1990*. Taylor and Francis.

Dunlap, Riley E., and Rik Scarce. 1991. Poll trends: Environmental problems and protection. *Public Opinion Quarterly* 55: 651–672.

Dunlap, Riley E., and K. D. Van Liere. 1978. The new environmental paradigm: A proposed measuring instrument and preliminary results. *Journal of Environmental Education* 9: 10–19.

Dunlap, Riley E., George H. Gallup Jr., and Alex M. Gallup. 1992. *The Health of the Planet Survey*. George H. Gallup International Institute.

Dutton, W. H., James N. Danziger, and Kenneth L. Kraemer. 1980. Did the policy fail? The selective use of automated information in the policy-making process. In *Why Policies Succeed or Fail*, ed. H. Ingram et al. Sage.

Earth First! 1987. Earth First! introductory guide. *Earth First! Journal* handout.

Eckersley, Robyn. 1990. Habermas and green political thought: Two roads diverging. *Theory and Society* 19: 739–776.

Eckersley, Robyn. 1992a. The failed promise of critical theory. Reprinted in *Ecology*, ed. C. Merchant and R. Gottlieb (Humanities, 1994).

Eckersley, Robyn. 1992b. *Environmentalism and Political Theory: Toward an Ecocentric Approach*. State University of New York Press.

Eckersley, Robyn. 1999. The discourse ethic and the problem of representing nature. *Environmental Politics* 8, no. 2: 24–49.

Eddy, J. Arthur. 1909, Conservation Alarms. Bulletin no. 8, National Public Domain League.

Eden, S. E. 1993. Individual environmental responsibility and its role in public environmentalism. *Environment and Planning A* 25: 1743–1758.

Eder, Klaus. 1985. The new social movements: Moral crusades, political pressure groups, or social movements? *Social Research* 52, no. 4: 869–900.

Eder, Klaus. 1992. Politics and culture: On the sociocultural analysis of political participation. In *Cultural-Political Interventions in the Unfinished Project of Enlightenment*, ed. A. Honneth et al. MIT Press.

Eder, Klaus. 1993. *The New Politics of Class: Social Movements and Cultural Dynamics in Advanced Societies*. Sage.

Eder, Klaus. 1996a. The institutionalization of environmentalism: Ecological discourse and the second transformation of the public sphere. In *Risk, Environment and Modernity*, ed. S. Lash et al. Sage.

Eder, Klaus. 1996b. *The Social Construction of Nature*. Sage.

Ehrlich, Paul. 1968. *The Population Bomb*. Ballantine.

Ehrlich, Paul R. 1986. *The Machinery of Nature*. Simon and Schuster.

Ehrlich, Paul R., and John P. Holdren. 1971. Impact of population growth. *Science* 171: 1212–1217.

Elkins, Stephan. 1990. The politics of mystical ecology. *Telos* 82: 52–70.

Ellefson, Paul V. 1992. *Forest Resources Policy: Process, Participants, and Programs*. McGraw-Hill.

Elliot, Robert, and Arran Gare, eds. 1983. *Environmental Philosophy*. Pennsylvania State University Press.

Ely, John. 1983 The Greens: Ecology and the promise of radical democracy. *Radical America* 17, March–June: 23–34.

Ely, John. 1992. The politics of "civil society." *Telos* 93: 173–191.

Emirbayer, Mustafa, and Jeff Goodwin. 1994. Network analysis, culture, and the problem of agency. *American Journal of Sociology* 99, no. 6: 1411–1454.

Environmental Careers Organization. 1992. *Beyond the Green: Redefining and Diversifying the Environmental Movement*.

Environmental Grantmakers Association. 1992. Fall Retreat Transcripts.

Enzensberger, H. M. 1979. A Critique of Political Ecology. In *Political Ecology*, ed. A. Cockburn and J. Ridgeway. Times Books.

Epstein, Richard A. 1985. *Takings: Private Property and the Power of Eminent Domain*. Harvard University Press.

Etzioni, A. 1976 *Social Problems*. Prentice-Hall.

Evangelical Lutheran Church in America. 1990. Basis for Our Caring. Reprinted in *This Sacred Earth*, ed. R. Gottlieb (Routledge, 1996).

Evernden, Neil. 1992a. Ecology in conservation and conversation. In *After Earth Day*, ed. M. Oelschlaeger. University of North Texas Press.

Evernden, Neil. 1992b. *The Social Creation of Nature*. Johns Hopkins University Press.

Ewen, Stuart. 1976. *Captains of Consciousness: Advertising and the Social Roots of the Consumer Culture*. McGraw-Hill.

Eyerman, R., and A. Jamison. 1991. *Social Movements: A Cognitive Approach*. Pennsylvania State University Press.

Fabrizio, Ray, Edith Karas, and Ruth Menmuir. 1970. *The Rhetoric of NO*. Holt, Rinehart and Winston.

FAITH-MAN-NATURE Group. 1967. Christians and the Good Earth: Addresses and discussions at the Third National Conference of the FAITH-MAN-NATURE GROUP.

Feenberg, Andrew. 1991. *Critical Theory of Technology*. Oxford University Press.

Ferris, Deeohn, and David Hahn-Baker. 1995. Environmentalists and Environmental Justice Policy. In *Environmental Justice*, ed. B. Bryant. Island.

Fireman, B., W. A. Gamson, S. Rytina, and B. Taylor. 1979. Encounters with unjust authority. *Social Movements, Conflict, and Change* 2: 1–33.

FNPCELS (First National People of Color Environmental Leadership Summit). 1991. Principles of Environmental Justice. Reprinted in *Ecology*, ed. C. Merchant and R. Gottlieb (Humanities, 1994).

Fischer, Frank. 1990. *Technocracy and the Politics of Expertise*. Sage.

Fischer, Frank. 1992. The Greening of Risk Assessment: Towards a Participatory Approach. In *Business and the Environment*, ed. D. Smith. Paul Chapman.

Fischer, Frank. 1993. Citizen participation and the democratization of policy expertise: From theoretical inquiry to practical cases. *Policy Sciences* 26: 165–187.

Foreman, Dave. 1991. *Confessions of an Eco-Warrior*. Harmony.

Foreman, Dave, and Bill Haywood. 1985. *Ecodefense: A Field Guide to Monkeywrenching*. Ned Ludd Books.

Forester, J., ed. 1985. *Critical Theory and Public Life*. MIT Press.

Foss, Phillip O. 1960. *Politics and Grass: The Administration of Grazing on the Public Domain*. Greenwood.

Foucault, M. 1972. *The Archaeology of Knowledge and the Discourse on Language*. Pantheon.

Fox, Matthew. 1983. *Original Blessing: A Primer in Creation Spirituality*. Bear.

Fox, Stephen. 1981. *The American Conservation Movement: John Muir and His Legacy*. University of Wisconsin Press.

Fox, Stephen. 1984. We want no straddlers. *Wilderness*, winter: 5–19.

Fox, Warwick. 1989. The deep ecology–ecofeminism debate and its parallels. *Environmental Ethics* 11, spring: 5–25.

Frame, Randy. 1996. Greening of the gospel? *Christianity Today*, November 11: 82–86.

Frankel, Boris. 1982. On the state of the state: Marxist theories of the state after Leninism. In *Classes, Power, and Conflict*, ed. A. Giddens et al. University of California Press.

Fredrickson, George M. 1965. *The Inner Civil War: Northern Intellectuals and the Crisis of the Union*. Harper and Row.

French, Roderick S. 1980. Ecological humanism a contradiction in terms? The philosophical foundations of the humanities under attack. In *Ecological Consciousness*, ed. R. Schultz and J. Hughes. University Press of America.

Freudenberg, Nicholas. 1984. *NOT in our Backyards! Community Action for Health and the Environment*. Monthly Review Press.

Freudenberg, Nicholas, and Carol Steinsapir. 1992. Not in our backyards: The grassroots environmental movement. In *The U.S. Environmental Movement, 1970–1990*, ed. R. Dunlap and A. Mertig. Taylor and Francis.

Friedman, John. 1992. *Empowerment: The Politics of Alternative Development.* Blackwell.

Friedman, Milton. 1962. *Capitalism and Freedom.* University of Chicago Press.

Friends of the Land for Soil and Water conservation. 1940. Transcript of Proceedings at the Organization Meeting.

Gadamer, Hans-Georg. 1984. *Truth and Method.* Crossroads.

Gale, Richard. 1986. Social Movements and the State: The Environmental Movement, Countermovement, and Government Agencies. *Sociological Perspectives* 9, no. 2: 179–199.

Gale, Richard. 1990. *Encyclopedia of Associations.* Gale.

Gamson, W. A. 1975. *The Strategy of Social Protest.* Dorsey.

Gamson, W. A. 1985. Goffman's legacy to political sociology. *Theory and Society* 14, no. 5: 605–622.

Gamson, W. A. 1988. Political Discourse and Collective Action. In *International Social Movement Research*, ed. B. Klandermans et al. JAI.

Gamson, W. A. 1991. Commitment and agency in social movements. *Sociological Forum* 6, no. 1: 27–50.

Gamson, W. A. 1992. The Social Psychology of Collective Action. In *Frontiers in Social Movement Theory*, ed. A. Morris and C. Mueller. Yale University Press.

Gamson, W. A., and Gadi Wolfsfeld. 1993. Movements and media as interacting systems. *Annals of the American Academy of Political and Social Science* no. 528: 114–126.

Garretson, Martin S. 1934, *A Short History of the American Bison.* American Bison Society.

Gedicks, Al. 1993. *The New Resource Wars: Native and Environmental Struggles Against Multinational Corporations.* South End.

Gerhards, Jürgen, and Dieter Rucht. 1992. Mesomobilization: Organizing and framing in two protest campaigns in West Germany. *American Journal of Sociology* 98, no. 3: 555–595.

Gibbs, Lois Marie. 1982. *Love Canal: My Story.* State University of New York Press.

Gibbs, Lois Marie, and Karen J. Stults. 1988. On grassroots environmentalism. In *Crossroads*, ed. P. Borecelli. Island.

Giddens, A. 1985. Jürgen Habermas. In *The Return of Grand Theory in the Human Sciences*, ed. Q. Skinner. Cambridge University Press.

Giddens, A. 1994. What can radical politics look like today? Talk presented at the International Sociological Association conference, Bielefield, Germany.

Gifford, Bill. 1991. Report card time! Doling out the marks on 14 leading environmental groups. *Outside Magazine*, December: 38–39.

Gilbert, Frederick F., and Donald G. Dodds. 1987. *The Philosophy and Practice of Wildlife Management.* Krieger.

Gilpin, William. 1846. *Mission of the North American People, Geographical, Social and Political* (Lippincott, 1973).

Godwin, R. Kenneth, and Robert C. Mitchell. 1984. The implications of direct mail for political organizations. *Social Science Quarterly* 65: 829–839.

Goldblatt, David. 1996. *Social Theory and the Environment*. Westview.

Goodwin, L. 1975. *Can Social Science Help Resolve National Problems?* Free Press.

Gottlieb, Alan M., ed. 1989. *The Wise Use Agenda*. Free Enterprise Press.

Gottlieb, Robert. 1993. *Forcing the Spring: The Transformation of the American Environmental Movement*. Island.

Gottlieb, Robert, and Helen Ingram. 1988. The new environmentalists. *Progressive*, August: 14–15.

Gottlieb, Roger S. 1995. Spiritual deep ecology and the left: An attempt at reconciliation. Reprinted in Gottlieb, *This Sacred Earth* (Routledge, 1996).

Gottlieb, Roger S. 1996. *This Sacred Earth*. Routledge.

Gould, K. A., A. Schnaiberg, and A. S. Weinberg, A. S. 1996. *Local Environmental Struggles: Citizen Activism in the Treadmill of Production*. Cambridge University Press.

Goulet, Denis. 1995. Authentic development: Is it sustainable? In *A Sustainable World*, ed. T. Trzyna. International Union for the Conservation of Nature.

Gowdy, John M. 1998. Biophysical limits to industrialization: Prospects for the twenty-first century. In *The Coming Age of Scarcity*, ed. M. Dobkowski and I. Wallimann. Syracuse University Press.

Graf, William L. 1990. *Wilderness Preservation and the Sagebrush Rebellions*. Rowman and Littlefield.

Graham, John D., and March Sadowitz. 1994. Superfund reform: Reducing risk through community choice. *Issues in Science and Technology*, summer: 35–40.

Gramsci, A. 1971. *Selections from the Prison Notebooks*. International.

Greene, S. G. 1995. Surviving in a grassroots environment. *Chronicle of Philanthropy* 8, no. 4: 1, 12–14.

Greimas, A. J. 1987. *On Meaning: Selected Writings in Semiotic Theory*. University of Minnesota Press.

Griffin, Leland M. 1966. A dramatistic theory of the rhetoric of movements. Reprinted in *Critical Responses to Kenneth Burke: 1924–1966*, ed. W. Rueckert (University of Minnesota Press, 1969).

Griffin, Leland M. 1980. On studying movements. *Central States Speech Journal* 31, no. 4: 225–231.

Grinnell, George Bird. 1910. *Brief History of the Boone and Crockett Club*. Forest and Stream.

Grinnell, George Bird, and Charles Sheldon, eds. 1895. *Hunting and Conservation: The Book of the Boone and Crockett Club*. Yale University (reprint: Arno, 1970).

Grossman, Karl. 1994. The People of Color Environmental Summit. In *Unequal Protection: Environmental Justice and Communities of Color*, ed. R. Bullard. Sierra Club.

Grove, Richard H. 1990. Colonial conservation, ecological hegemony and popular resistance: Towards a global synthesis. In *Imperialism and the Natural World*, ed. J. MacKenzie. St. Martin's Press.

Grove, Richard H. 1992. Origins of Western environmentalism. *Scientific American*, July: 42–47.

Grumbine, R. E. 1994. Wildness, wise use, and sustainable development. *Environmental Ethics* 16: 227–249.

Grumbine, R. E. 1997. Using biodiversity as a justification for nature protection in the U.S. *Wild Earth*, winter 1996–97: 71–80.

Habermas, Jürgen. 1962. *The Structural Transformation of the Public Sphere: An Inquiry into a Category of Bourgeois Society* (MIT Press, 1989).

Habermas, Jürgen. 1970. *Toward a Rational Society*. Beacon.

Habermas, Jürgen. 1971. *Knowledge and Human Interests*. Beacon.

Habermas, Jürgen. 1973. *Theory and Practice*. Beacon.

Habermas, Jürgen. 1975. *Legitimation Crisis*. Beacon.

Habermas, Jürgen. 1979. *Communication and the Evolution of Society*. Beacon.

Habermas, Jürgen. 1983a. Hermeneutics and critical theory. Presented at Bryn Mawr College.

Habermas, Jürgen. 1983b. *Philosophical-Political Profiles*. MIT Press.

Habermas, Jürgen. 1984. *The Theory of Communicative Action, Volume One, Reason and the Rationalization of Society*. Beacon.

Habermas, Jürgen, ed. 1985a. *Observations on the Spiritual Situation of the Age*. MIT Press.

Habermas, Jürgen. 1985b. Neoconservative culture criticism the United States and West Germany: An intellectual movement in two political cultures. In *Habermas and Modernity*, ed. R. Bernstein. MIT Press.

Habermas, Jürgen. 1986a. The new obscurity: The crisis of the welfare state and the exhaustion of utopian energies. *Philosophy and Social Criticism* 11, no. 2: 385–410.

Habermas, Jürgen. 1986b. Life-forms, morality and the task of the philosopher. In *Autonomy and Solidarity*, ed. P. Dews. Verso.

Habermas, Jürgen. 1987a. *The Theory of Communicative Action, Volume Two Lifeworld and System: A Critique of Functionalist Reason*. Beacon.

Habermas, Jürgen. 1987b. *The Philosophical Discourse of Modernity*. MIT Press.

Habermas, Jürgen. 1989a. The Public Sphere. In *Critical Theory and Society*, ed. S. Bronner and D. Kellner. Routledge.

Habermas, Jürgen. 1989b. *The New Conservatism: Cultural Criticism and the Historians' Debate.* MIT Press.

Habermas, Jürgen. 1989c. *The Structural Transformation of the Public Sphere: An Inquiry into a Category of Bourgeois Society.* MIT Press.

Habermas, Jürgen. 1990. *Moral Consciousness and Communicative Action.* MIT Press.

Habermas, Jürgen. 1991. A reply. In *Communicative Action,* ed. A. Honneth and H. Joas. MIT Press.

Habermas, Jürgen. 1992a. Further Reflections on the Public Sphere. In *Habermas and the Public Sphere,* ed. C. Calhoun. MIT Press.

Habermas, Jürgen. 1992b. *Postmetaphysical Thinking.* MIT Press.

Habermas, Jürgen. 1993. *Justification and Application: Remarks on Discourse Ethics.* MIT Press.

Habermas, Jürgen. 1994. *The Past as Future.* University of Nebraska Press.

Habermas, Jürgen. 1996. *Between Facts and Norms: Contributions to a Discourse Theory of Law and Democracy.* MIT Press.

Habermas, Jürgen. 1998a. *On the Pragmatics of Communication.* MIT Press.

Habermas, Jürgen. 1998b. *The Inclusion of the Other: Studies in Political Theory.* MIT Press.

Hallock, Charles. 1878. *Hallock's American Club List and Sportsman's Glossary.* Forest and Stream.

Hamilton, Alice. 1925. *Industrial Poisons in the United States.* Macmillan.

Hannigan, John A. 1995. *Environmental Sociology: A Social Constructionist Perspective.* Routledge.

Hardin, Garett. 1968. The Tragedy of the Commons. *Science* 162: 1243–1248.

Hare, Nathan. 1970. Black ecology. *The Black Scholar* 1, no. 6: 2–8.

Harjo, Susan Shown. 1992. Statement of Vision Toward the Next 500 Years. Morning Star Foundation, Washington.

Harre, Rom, Jens Brockmeier, and Peter Muhlhausler. 1999. *Greenspeak: A Study of Environmental Discourse.* Sage.

Harvard School of Public Health. 1937. *The Environment and Its Effect Upon Man.*

Hatch, Sen. Orrin G. 1981. The stewardship of the public domain. In *Agenda for the '80s: A New Federal Land Policy.* LASER.

Hayes, Michael T. 1978. The semi-sovereign pressure groups: a critique of current theory and an alternative typology. *Journal of Politics* 40.

Hayes, Michael T. 1986. The new group universe. In *Interest Group Politics,* ed. A. Cigler and B. Loomis. Congressional Quarterly Press.

Hays, Samuel P. 1959. *Conservation and the Gospel of Efficiency: The Progressive Conservation Movement, 1890–1920.* Athenaeum.

Hays, Samuel P. 1981. The structure of environmental politics since World War II. *Journal of Social History* 14, no. 4: 719–738.

Hays, Samuel P. 1985. From conservation to environment: Environmental politics in the U.S. Since World War II. In *Environmental History*, ed. K. Bailes. University Press of America.

Hays, Samuel P. 1987a. *Beauty, Health, and Permanence: Environmental Politics in the United States, 1955–1988*. Cambridge University Press.

Hays, Samuel P. 1987b. The politics of environmental administration. In *The New American State*, ed. L. Galambos. Johns Hopkins University Press.

Hays, Samuel P. 1992. environmental political culture and environmental political development: An analysis of legislative voting, 1971–1989. *Environmental History Review* 16, no. 2: 1–22.

Hazelrigg, Lawrence. 1995. *Cultures of Nature: An Essay on the Production of Nature*. University Press of Florida.

Hedstrom, Peter. 1994. Contagious collectivities: On the spatial diffusion of Swedish trade unions, 1890–1940. *American Journal of Sociology* 99, no. 5: 1157–1179.

Held, David. 1980. *Introduction to Critical Theory*. University of California Press.

Held, David. 1982. Crisis Tendencies, Legitimation and the State. In *Habermas: Critical Debates*, ed. J. Thompson and D. Held. MIT Press.

Held, David, and J. Krieger. 1983 Accumulation, legitimation and the state: The ideas of Claus Offe and Jürgen Habermas. In *States and Societies*, ed. D. Held. New York University Press.

Helvarg, David. 1994. *The War against the Greens: The "Wise-Use" Movement, the New Right, and Anti-Environmental Violence*. Sierra Club.

Herbert, Henry William. 1849, *Frank Forester's Field Sports of the United States and British Provinces of North America*. American News Company.

Hermans, J. M., H. Kempen, and R. van Loon. 1992. The dialogical self. *American Psychologist*, January: 23–33.

Herndl, Carl G., and Stuart C. Brown. 1996. Rhetorical criticism and the environment. In *Green Culture*, ed. C. Herndl and S. Brown. University of Wisconsin Press.

Highlander Research and Education Center. 1992. *Environment and Development in the USA: A Grassroots Report for UNCED*.

Hilgartner, Stephen, and Charles C. Bosk. 1988. The rise and fall of social problems: A public arenas model. *American Journal of Sociology* 94, no. 1: 53–78.

Holdren, John P. 1997. Press conference on climate change, June 18.

Holub, Robert C. 1991. *Jürgen Habermas: Critic in the Public Sphere*. Routledge.

Homer-Dixon, Thomas F., Jeffrey H. Boutwell, and George W. Rathjens. 1993. Environmental change and violent conflict. *Scientific American* 268, no. 2: 38–45.

Honneth, Axel, and Hans Joas, eds. 1991. *Communicative Action: Essays on Jürgen Habermas's The Theory of Communicative Action*. MIT Press.

Horkheimer, Max. 1937. Postscript to traditional and critical theory. In *Critical Theory*, ed. M. Horkheimer (Continuum, 1992).

Hough, F. B. 1873. On the duty of governments in the preservation of forests. In *Proceedings of the American Association for the Advancement of Science, Twenty Second Meeting.*

Hough, Romeyn B. 1913. The incipiency of the forestry movement in America. *American Forestry* 3: 547–550.

Hoy, David Couzens, and Thomas McCarthy. 1994. *Critical Theory*. Blackwell.

Hoy, Suellen M. 1980. Municipal housekeeping: the role of women in improving urban sanitation practices 1880–1917. In *Pollution and Reform in American Cities, 1870–1930*, ed. M. Melosi. University of Texas Press.

Hughs, H. Stuart. 1988. *Sophisticated Rebels: The Political Culture of European Dissent 1968–1987*. Harvard University Press.

Humboldt, Alexander von. 1849. *Aspects of Nature in Different Lands and Different Climates*. Wilson and Ogilby.

Humboldt, Alexander von. 1850a. *Cosmos: A sketch of a physical description of the universe*. Bohn.

Humboldt, Alexander von. 1850b. *Views of nature, or, Contemplations on the sublime phenomena of creation*. Bohn.

Hummel, Richard. 1994. *Hunting and Fishing for Sport: Commerce, Controversy, Popular Culture*. Bowling Green State University.

Humphrey, Craig R., and Frederick R. Buttel. 1982. *Environment, Energy and Society*. Wadsworth.

Huth, Hans. 1948. Yosemite: The story of an idea. *Sierra Club Bulletin*, March: 47–48.

Huth, Hans. 1957, *Nature and the American: Three Centuries of Changing Attitudes*. University of California Press.

Inglehart, Ronald. 1986. Intergenerational changes in politics and culture: The shift from materialist to postmaterialist value priorities. *Research in Political Sociology* 2: 81–105.

Ingram, Helen M., and Mann, Dean E. 1989. Interest Groups and Environmental Policy. In *Environmental Politics and Policy: Theories and Evidence*, ed. J. Lester. Duke University Press.

International Physician. 1997. Letter on global climate change and human health, June 23.

IPCC (Intergovernmental Panel on Climate Change). 1995. *Climate Change 1995, the IPCC Synthesis*. United Nations.

Issel, Bernardo. 1999. Crony Environmentalism. NonProfit Accountability Project, Washington, D.C.

IUCN (International Union for the Conservation of Nature). 1996. IUCN Red List of Threatened Animals.

IUCN. 1997. IUCN Red List of Threatened Plants.

Jablonski, Carol J. 1980. Promoting radical change in the Roman Catholic Church: Rhetorical requirements, problems, and strategies of the American bishops. *Central States Speech Journal* 31, no. 4: 282–289.

Jaeger, Carlo C. 1992. Regional approaches to global climatic risks. Presented at symposium on Current Developments in Environmental Sociology, Amsterdam.

Jaeger, Carlo C. 1994. *Taming the Dragon: Transforming Economic Institutions in the Face of Global Change*. Gordon and Breach.

Janicke, Martin. 1990. *State Failure: The Impotence of Politics in Industrial Society*. Pennsylvania State University Press.

Jenkins, J. Craig. 1983. Resource Mobilization Theory and the Study of Social Movements. *Annual Review of Sociology* 9.

Jenkins, J. Craig. 1987. Nonprofit organizations and policy advocacy. In *The Nonprofit Sector*, ed. W. Powell. Yale University Press.

Jenkins, J. Craig. 1989. Social Movement Philanthropy and American Democracy. In *Philanthropic Giving*, ed. R. Magat. Oxford University Press.

Jenkins, J. Craig. 1996. Channeling social protest: Foundation patronage of contemporary social movements. In *Private Action and the Public Good*, ed. W. Powell and E. Clemens. Yale University Press.

Jenkins, J. Craig, and Craig M. Eckert. 1986. Channeling black insurgency: Elite patronage and professional social movement organizations in the development of the black movement. *American Sociological Review* 51: 812–829.

Jenkins, J. Craig, and A. Halcli. 1994. The Institutionalization of Social Movement Philanthropy: Three Decades of Foundation Giving. 1960–1990. Unpublished manuscript.

Jenkins, J. Craig, and A. Halcli. 1996. Grassrooting the system? The development and impact of social movement philanthropy. 1953–1990. Presented at conference on Philanthropic Foundations in History: Needs and Opportunities, New York University.

Jessop, Bob. 1982. *The Capitalist State*. New York University Press.

John Paul II. 1990. The ecological crisis: A common responsibility. Reprinted in *This Sacred Earth*, ed. R. Gottlieb (Routledge, 1996).

Jones, Holway R. 1965. *John Muir and the Sierra Club: The Battle for Yosemite*. Sierra Club.

Jordan, Charles, and Donald Snow. 1992., Diversification, Minorities, and the Mainstream Environmental Movement. In *Voices from the Environmental Movement*, ed. D. Snow and P. Noonan. Island.

Kaase, Max. 1990. Social Movements and Political Innovation. In *Challenging the Political Order*, ed. R. Dalton and M. Kuechler. Oxford University Press.

Kastenholz, Hans G., and Karl-Heinz Erdmann. 1992. Positive social behavior and the environmental crisis. *The Environmentalist* 12, no. 3: 181–186.

Keane, John, ed. 1988. *Civil Society and the State: New European Perspectives.* Verso.

Kearns, Laurel. 1996. Saving the creation: Christian environmentalism in the United States. *Sociology of Religion* 57: 1.

Kelber, Mim. 1994. The Women's Environment and Development Organization. *Environment* 36, no. 8: 43–45.

Killingsworth, M. J, and Jacqueline S. Palmer. 1992. *Ecospeak: Rhetoric and Environmental Politics in America.* Southern Illinois University Press.

Killingsworth, M. J, and Jacqueline S. Palmer. 1996. Millennial Ecology: The Apocalyptic Narrative from Silent Spring to Global Warming. In *Green Culture,* ed. C. Herndl and S. Brown. University of Wisconsin Press.

Klandermans, Bert. 1988. The formation and mobilization of consensus. In *International Social Movement Research,* ed. B. Klandermans et al. JAI.

Klandermans, Bert. 1991. New Social Movements and Resource Mobilization: The European and the American Approach Revisited. In *Research on Social Movements,* ed. D. Rucht. Westview.

Klandermans, Bert. 1992. The Social Construction of Protest and Multiorganizational Fields. In *Frontiers in Social Movement Theory,* ed. A. Morris and C. Mueller. Yale University Press.

Knoke, David. 1990. *Organizing for Collective Action: The Political Economies of Associations.* Aldine de Gruyter.

Knox, Margaret L. 1990. The wise use guys. *Buzzworm* 2, no. 6: 30–36.

Krauss, Celene. 1993a. Blue Collar Women and Toxic-Waste Protests. In *Toxic Struggles,* ed. R. Hofrichter. New Society.

Krauss, Celene. 1993b. Women and toxic waste protests: race, class and gender as resources of resistance. *Qualitative Sociology* 16, no. 3: 247–262.

Krauss, Celene. 1994. Women of color on the front line. In *Unequal Protection,* ed. R Bullard. Sierra Club.

Krupp, Frederic D. 1986. New environmentalism factors in economic needs. *Wall Street Journal,* November 20.

Kuhn, T. 1962. *The Structure of Scientific Revolutions.* University of Chicago Press.

Kurtz, L. R. 1983. The Politics of heresy. *American Journal of Sociology* 88, no. 6: 1085–1115.

Laclau, Ernesto, and Chantal Mouffe. 1985. *Hegemony and Socialist Strategy: Towards A Radical Democratic Politics.* Verso.

LaDuke, Winona. 1994. From resistance to regeneration. In *Ecology,* ed. C. Merchant and R. Gottlieb. Humanities.

Lamb, John. 1913. The states' rights question. *American Forestry,* March: 172–190.

Landau, Misia. 1991. *Narratives of Human Evolution.* Yale University Press.

Langton, Stuart. 1984. Networking and the environmental movement. In *Environmental Leadership*, ed. S. Langton. Heath.

Lanier-Graham, Susan D. 1991. *The Nature Directory: A Guide to Environmental Organizations*. Walker.

Lapp, David. 1992. Reverend Jesse Jackson—Fighting for the right to breathe free, *E: The Environmental Magazine* 3 (May-June): 10–18.

LASER. 1981. *Agenda for the '80s: A New Federal Land Policy*.

Laumann, Edward O., and David Knoke. 1987. *The Organizational State: Social Choice in National Policy Domains*. University of Wisconsin Press.

Lave, L. B., and E. P. Seskin. 1970. Air pollution and human health. *Science* 169: 728.

Lavelle, Marianne, and Marcia A. Coyle. 1993. Unequal Protection: The Racial Divide in Environmental Law. In *Toxic Struggles*, ed. R. Hofrichter. New Society.

Lee, Charles. 1993. Beyond Toxic Wastes and Race. In *Confronting Environmental Racism*, ed. R. Bullard. South End.

Leopold, Aldo. 1933. *Game Management*. Scribner.

Leopold, Aldo. 1949, *Sand County Almanac*. Ballantine.

Lessl, Thomas M. 1989. The priestly voice. *Quarterly Journal of Speech* 75: 183–197.

Lipset, Seymour M., M. Trow, and J. Coleman. 1956. *Union Democracy*. Free Press.

Livesay, Jess. 1994. Post-Marxist theories of civil society. *Current Perspectives in Social Theory* 14: 101–134.

Lo, Clarence Y. H. 1982. Countermovements and conservative movements in the contemporary U.S. *Annual Review of Sociology* 8: 107–134.

Lofland, John, Mary Anna, Colwell, and Victoria Johnson. 1989. Social movement strategies as theories of social change. Presented at 1989 annual Meeting of American Sociological Association, San Francisco.

Lowe, George D., and Thomas K. Pinhey. 1982. Rural-urban differences in support for environmental protection. *Rural Sociology* 47, no. 1: 114–128.

Lowe, Philip, and Jane Goyder. 1983. *Environmental Groups in Politics*. Allen and Unwin.

Lowry, Robert C. 1993. Interest, ideology, and market penetration in a social movement industry: The case of environmental citizen groups. Presented at Annual Meeting of American Political Science Association, Washington.

Lubin, David M. 1994. *Picturing a Nation: Art and Social Change in Nineteenth-Century America*. Yale University Press.

Lukacs, Georg. 1968. *History and Class Consciousness: Studies in Marxist Dialectics*. MIT Press.

Luke, Timothy. 1988. The dreams of deep ecology. *Telos* 76: 65–92.

Luke, Timothy. 1991. Community and ecology. *Telos* 88: 69–79.

Luke, Timothy. 1994. Ecological politics and local struggles: Earth First! as an environmental resistance movement. *Current Perspectives in Social Theory* 14: 241–267.

Luke, Timothy, and Stephen K. White. 1985. Critical theory, the informational revolution, and an ecological path to modernity. In *Critical Theory and Public Life*, ed. J. Forster. MIT Press.

Lynne, J. 1990. Bio-rhetorics: Moralizing the life sciences. In *The Rhetorical Turn*, ed. H. Simons. University of Chicago Press.

MacKenzie, John M. 1988. *The Empire of Nature: Hunting, Conservation and British Imperialism.* St. Martin's Press.

Madison, Gary B. 1981. *The Phenomenology of Merleau Ponty: A Search for the Limits of Consciousness.* Ohio University Press.

Manes, Christopher. 1990. *Green Rage.* Little, Brown.

Mann, M. 1970. The social cohesion of liberal democracy. *American Sociological Review* 35, no. 3: 432–439.

Manning, Roger B. 1993. *Hunters and Poachers: A Social and Cultural History of Unlawful Hunting in England, 1485–1640.* Clarendon.

Marcus, Melvin G. 1985. Environmental policies in the United States. In *Environmental Policies*, ed. C. Park. Croom Helm.

Marcuse, Herbert. 1972. *Counter-Revolution and Revolt.* Beacon.

Marcuse, Peter. 1974. Conservation for whom? In *Environmental Quality and Social Justice in Urban America*, ed. J. Smith. Conservation Foundation.

Marsh, G. P. 1864. *Man and Nature.* Scribner.

Marshall, Robert. 1930. The problem of the wilderness. *Mountain Magazine*, May 1930.

Marshall, Robert. 1933. *The People's Forests.* H. Smith and R. Haas.

Martel, N., and B. Holman. 1994. Inside the environmental groups. *Outside Magazine*, March: 64–73.

Martin, Russel. 1989. *A Story That Stands Like a Dam: Glen Canyon and the Struggle for the Soul of the West.* Holt.

Marx, Gary T., and James L. Wood. 1975. *Strands of Theory and Research in Collective Behavior*, volume 1. Annual Reviews Inc.

Marx, Karl. 1843. Letters from the Franco-German yearbooks. In *Early Writings* (Vintage, 1975).

Marx, Karl. 1867. *Capital* (International Publishers, 1967).

Mason, Augustus L. 1896. *American Pioneer History.* National Book Company.

Mathews, Freya. 1994. Ecofeminism and deep ecology. In *Ecology*, ed. C. Merchant and R. Gottlieb. Humanities.

Maughan, Ralph, and Douglas Nilson. 1993. What's old and what's new about the wise use movement. Presented at Western Social Science Association Convention.

Mauss, J. 1989. Beyond the illusion of social problems theory. *Perspectives on Social Problems* 1: 19–39.

Maynard, D. W., and T. P. Wilson. 1980. On the reification of social structure. *Current Perspectives in Social Theory* 1: 287–322.

McCarthy, John D., David W. Britt, and Mark Wolfson. 1991. The institutional channeling of social movements by the state in the United States. *Research in Social Movements, Conflicts and Change* 13: 45–76.

McCarthy, Thomas. 1978. *The Critical Theory of Jürgen Habermas*. MIT Press.

McCarthy, Thomas. 1991a. Complexity and democracy: or the seducements of systems theory. In *Communicative Action*, ed. A. Honneth and H. Joas. MIT Press.

McCarthy, Thomas. 1991b. *Ideals and Illusions: On Reconstruction and Deconstruction in Contemporary Critical Theory*. MIT Press.

McCarthy, Thomas. 1992. Practical discourse: On the relation of morality to politics. In *Habermas and the Public Sphere*, ed. C. Calhoun. MIT Press.

McCormick. John. 1989. *Reclaiming Paradise: The Global Environmental Movement*. Indiana University Press.

McDonald, Angus. 1941. Early American Soil Conservationists. Miscellaneous Publication no. 449, U.S. Department of Agriculture.

McGee, Michael Calvin. 1980a. Social movement: Phenomena or meaning? *Central States Speech Journal* 31, no. 4: 232–244.

McGee, Michael Calvin. 1980b. The ideograph, a link between rhetoric and ideology. *Quarterly Journal of Speech* 66: 1–16.

McHugh, Josh. 1994. Greenpeace's greenbacks. *Forbes*, November 21: 14.

McLoughlin, W. G. 1978. *Revivals, Awakenings, and Reform*. University of Chicago Press.

Mead, Georg Herbert. 1934. *Mind, Self, and Society: From the Standpoint of a Social Behaviorist* (University of Chicago Press, 1962).

Meadows, D. H. 1972. *The Limits to Growth*. Universe Books.

Meadows, D. H., D. L. Meadows, and Jorgen Randers. 1992. *Beyond the Limits: Confronting Global Collapse, Envisioning a Sustainable Future*. Chelsea Green.

Meeker, Joseph W. 1972. *The Comedy of Survival: Studies in Literary Ecology*. Scribner.

Melosi, Martin V. 1977. *Pragmatic Environmentalist: Sanitary Engineer George E. Waring Jr*. Public Works Historical Society.

Melosi, Martin V. 1981. *Garbage in the Cities: Refuse, Reform, and the Environment, 1880–1980*. Texas A&M University Press.

Melosi, Martin V. 1985. Environmental reform in the industrial cities: The civic response to pollution in the Progressive era. In *Environmental History*, ed. K. Bailes. University Press of America.

Melucci, Alberto. 1985. The symbolic challenge of contemporary movements. *Social Research* 52, no. 4: 789–816.

Melucci, Alberto. 1988. Getting involved: Identity and mobilization in social movements. *International Social Movement Research* 1: 329–348.

Merchant, Carolyn. 1980. *The Death of Nature, Women, Ecology, and the Scientific Revolution*. Harper and Row.

Merchant, Carolyn. 1992. *Radical Ecology: The Search for a Livable World*. Routledge.

Merchant, Carolyn, and Roger Gottlieb, eds. 1994. *Ecology*. Humanities.

Merchant, Carolyn. 1995. *Earthcare: Women and the Environment*. Routledge.

Meyer, David S., and Suzanne Staggenborg. 1996. Movements, countermovements, and the structure of political opportunity. *American Journal of Sociology* 101, no. 6: 1628–1645.

Michels, Robert. 1915. *Political Parties: A Sociological Study of the Oligarchical Tendencies of Modern Democracy* (Free Press, 1962).

Milbrath, Lester W. 1984. *Environmentalists: Vanguard for a New Society*. State University of New York Press.

Milbrath, Lester W. 1989. *Envisioning a Sustainable Society: Learning Our Way Out*. State University of New York Press.

Milgram, Stanley. 1974. *Obedience to Authority*. Harper and Row.

Miller, G., and James A. Holstein. 1989. On the sociology of social problems. *Perspectives on Social Problems* 1: 1–16.

Miller, Roberta Balstad. 1991. Social science and the challenge of global environmental change. *International Social Science Journal* 43, August: 609–617.

Miller, Vernice D. 1993. Building on Our Past, Planning for Our Future. In *Toxic Struggles*, ed. R. Hofrichter. New Society.

Mills, C. W. 1959, *The Sociological Imagination*. Oxford University Press.

Mitchell, Robert C. 1979. National environmental lobbies and the apparent illogic of collective action. In *Collective Decision Making*, ed. C. Russell. Johns Hopkins University Press.

Mitchell, Robert C. 1989. From Conservation to Environmental Movement: The Development of the Modern Environmental Lobbies. In *Government and Environmental Politics*, ed. M. Lacey. Woodrow Wilson Center Press.

Mitchell, Robert C., Angela G. Mertig, and Riley E. Dunlap. 1992. Twenty years of environmental mobilization: Trends among national environmental organizations. In *The U.S. Environmental Movement, 1970–1990*, ed. R. Dunlap and A. Mertig. Taylor and Francis.

Mohai, Paul. 1985. Public concern and elite involvement in environmental-conservation issues. *Social Science Quarterly* 66, no. 4: 820–838.

Mohai, Paul, and Ben W. Twight. 1987. age and environmentalism: An elaboration of the Buttel model using national survey evidence. *Social Science Quarterly* 68, no. 4: 798–815.

Mol, Arthur P. J., and Gert Spaargaren. 1993. Environment, Modernity and the Risk Society: The Apocalyptic Horizon of Environmental Reform. *International Sociology* 8, no. 4: 431–463.

Molotch, H. L., and D. Boden. 1985. Talking social structure. *American Sociological Review* 50, no. 3: 273–288.

Moore, Barrington. 1966. *Social Origins of Dictatorship and Democracy: Lord and Peasant in the Making of the Modern World*. Beacon.

Morgan, M. Granger. 1993. Risk analysis and management. *Scientific American* 269, no. 1: 32–41.

Morgan-Hubbard, Margaret. 1993. Small is beautiful. *Environmental Action*, summer: 16–17.

Morris, Aldon. 1992. Political consciousness and collective action. In *Frontiers in Social Movement Theory*, ed. A. Morris and C. Mueller. Yale University Press.

Morris, Aldon, and C. Herring. 1987. Theory and research in social movements: A critical review. In *Annual Review of Political Science*, ed. S. Long. Ablex.

Morris, Aldon, and Carol McClurg Muller, eds. 1992. *Frontiers in Social Movement Theory*. Yale University Press.

Morrison, Denton E. 1980. The soft, cutting edge of environmentalism: Why and how the appropriate technology notion is changing the movement. *Natural Resources Journal* 20: 275–298.

Morrison, Denton E., and Riley E. Dunlap. 1986. Environmentalism and elitism: A conceptual and empirical analysis. *Environmental Management* 10, no. 5: 581–589.

Morse, Edward S. 1905. The steam whistle: A menace to public health. Presented before Massachusetts Association of Boards of Health.

Muir, John. 1916., *The Story of My Boyhood and Youth and a Thousand-Mile Walk to the Gulf*. Houghton Mifflin.

Murphy, R. 1983. The struggle for scholarly recognition. *Theory and Society* 12, no. 5: 631–658.

Myers, Norman. 1996. The world's forests: Problems and potentials. *Environmental Conservation* 23, no. 2: 156–168.

Naess, Arne. 1972. The shallow and the deep, long-range ecology movement. A summary. *Inquiry* 16: 95–100.

Nash, R. 1967. *Wilderness and the American Mind*. Yale University Press.

Nash, R. 1985. Rounding Out the American Revolution: Ethical Extension and the New Environmentalism. In *Environmental History*, ed. K. Bailes. University Press of America.

Nash, R. 1989a. The greening of religion. Reprinted in *This Sacred Earth*, ed. R. Gottlieb (Routledge, 1996).

Nash, R. 1989b. *The Rights of Nature: A History of Environmental Ethics*. University of Wisconsin Press.

National Conservation Association. 1908. Statement of Principles.

National Environmental Justice Advisory Committee. 1996. *The Model Plan for Public Participation.* U.S. Environmental Protection Agency, Office of Environmental Justice.

National Wildlife Federation. National Wildlife Federation Conservation Directory (series).

Neidhardt, F., and D. Rucht. 1991. The Analysis of Social Movements: The state of the art and some perspectives for further research. In *Research on Social Movements*, ed. D. Rucht. Westview.

Nesbitt, W. H., and J. Reneau, eds. 1988. *Records of North American Big Game*, ninth edition. Boone and Crockett Club.

Neuzil, Mark, and William Kovarik. 1996. *Mass Media and Environmental Conflict: America's Green Crusades.* Sage.

New York Sanitary Association. 1859. Constitution and Bylaws.

New York State Association for the Protection of Fish and Game. 1884. *Twenty-Sixth Annual Convention.* Buffalo Audubon Club.

Nielsen, Waldemar A. 1972. *The Big Foundations.* Columbia University Press.

Nisbet, Robert. 1982. *Prejudices: A Philosophical Dictionary*, Harvard University Press.

NORC (National Opinion Research Center). 1994. *General Social Surveys Codebook.* Roper Center.

Norris, Christopher. 1982. *Deconstruction: Theory and Practice.* Methuen.

Norton. Bryan G. 1991. *Toward Unity among Environmentalists.* Oxford University Press

Norusis, Marija J. 1993. SPSS for Windows Base System User's Guide, Release 6. 0. SPSS Inc.

NRC (National Research Council). 1996. *Understanding Risk: Informing Decisions in a Democratic Society.*

NRC (National Research Council). 1992. *Global Environmental Change: Understanding the Human Dimensions.* National Academy Press.

Nunner-Winkler, G. 1992. Knowing and Wanting: On Moral Development in Early Childhood. In *Cultural-Political Interventions in the Unfinished Project of Enlightenment*, ed. A. Honneth et al. MIT Press.

Oberschall, A. 1978. Theories of social conflict. *Annual Review of Sociology* 4: 291–315.

O'Brien, Jim. 1983. Environmentalism as a mass movement: Historical notes. *Radical America* 17, March–June: 7–27.

O'Callaghan, Kate. 1992. Whose agenda for America? *Audubon Magazine*, September-October: 80–92.

O'Connor, James. 1973. *The Fiscal Crisis of the State.* Blackwell.

O'Connor, James. 1984. *Accumulation Crisis*. Blackwell.

O'Connor, James. 1987. *The Meaning of Crisis*. Blackwell.

Oelschlaeger, Max. 1991. *The Idea of Wilderness: From Prehistory to the Age of Ecology*. Yale University Press.

Oeslchlaeger, Max. 1992a. *The Wilderness Condition: Essays on Environment and Civilization*. Island.

Oelschlaeger, Max., ed. 1992b. *After Earth Day: Continuing the Conservation Effort*. University of North Texas Press.

Oelschlaeger, Max. 1994. *Caring for Creation: An Ecumenical Approach to the Environmental Crisis*. Yale University Press.

Offe, Claus. 1981. The social sciences: contract research or social movements. *Current Perspectives in Social Theory* 2: 31–38.

Offe, Claus. 1984. *Contradictions of the Welfare State*. MIT Press.

Offe, Claus. 1985a. *Disorganized Capitalism*. MIT Press.

Offe, Claus. 1985b. New social movements: Challenging the boundaries of institutional politics. *Social Research* 52, no. 4: 817–868.

Offe, Claus. 1990. Reflections on the institutional self-transformation of movement politics: A tentative stage model. In *Challenging the Political Order*, ed. R. Dalton and M. Kuechler. Oxford University Press.

Offe, Claus. 1992. Bindings, shackles, brakes: On self-limitation strategies. In *Cultural-Political Interventions in the Unfinished Project of Enlightenment*, ed. A. Honneth et al. MIT Press.

Offe, Claus, and Volker Ronge. 1982. Theses on the theory of the state. In *Classes, Power, and Conflict*, ed. A. Giddens and D. Held. University of California Press.

Olsen, Marvin E., Dora G. Lodwick, and Riley E. Dunlap. 1992. *Viewing the World Ecologically*. Westview.

Opie, John. 1986. The environment and the frontier. In *American Frontier and Western Issues*, ed. R. Nichols. Greenwood.

O'Riordan, Timothy. 1979. Public interest environmental groups in the United States and Britain. *American Studies* 13, no. 3: 409–438.

Ortner, Sherry. 1974. Is female to male as nature is to culture? In *Women, Culture, and Society*, ed. M. Rosaldo and L. Lamphere. Stanford University Press.

Osborn, Fairfield. 1948. *Our Plundered Planet*. Grosset & Dunlap.

Ostheimer, John M., and Leonard G. Ritt. 1976. *Environment, Energy, and Black Americans*. Sage.

Ostrander, Susan A. 1995. *Money for Change: Social Movement Philanthropy at Haymarket People's Fund*. Temple University Press.

Outhwaite, William, and Tom Bottomore. 1993. *Twentieth Century Social Thought*. Blackwell.

Owen, A. L. R. 1983. *Conservation Under F. D. R.* Praeger.

Paehlke, Robert. 1988. Democracy, Bureaucracy, and Environmentalism. *Environmental Ethics* 10, no. 2: 291–308.

Paehlke, Robert. 1989. *Environmentalism and the Future of Progressive Politics.* Yale University Press.

Palmer, T. S. 1912. *Chronology and Index of the More Important Events in American Game Protection 1776–1911.* Government Printing Office.

Parkin, Sara. 1989. *Green Parties: An International Guide.* Heretic Books.

Parsons, Talcott. 1971. *The System of Modern Societies.* Prentice-Hall.

Pateman, Carole. 1970. *Participation and Democratic Theory.* Cambridge University Press.

Pedhazur, Elazar J. 1982. *Multiple Regression in Behavioral Research.* Holt, Rinehart and Winston.

Pell, Eve. 1990. Buying In. *Mother Jones,* April-May: 25.

Pepper, S. 1942 *World Hypothesis.* University of California Press.

Perreault, William D., Jr., and Laurence E. Leigh. 1989. Reliability of nominal data based on qualitative judgements. *Journal of Marketing Research* 26, May: 135–148.

Petulla, Joseph M. 1987. *Environmental Protection in the United States.* San Francisco Study Center.

Phillips, John C., and Lewis Webb Hill. 1930, *Classics of the American Shooting Field: A Mixed Bag for the Kindly Sportsman 1783–1926.* Houghton Mifflin.

Phillips, Susan D. 1991. Meaning and structure in social movements: Mapping the network of national Canadian women's organizations. *Canadian Journal of Political Science* 24, no. 4: 755–782.

Pichardo, Nelson A. 1995., The power elite and elite-driven countermovements: The Associated Farmers of California during the 1930s. *Sociological Forum* 10, no. 1: 21–50.

Pinchot, Gifford. 1910. The fight for conservation. In *American Environmentalism,* ed. D. Worster. Wiley.

Ponting, Clive. 1992. *A Green History of the World, The Environment and the Collapse of Great Civilizations.* St. Martin's Press.

Popper, Karl R. 1961: 1976. The logic of the social sciences. In *The Positivist Dispute in German Sociology,* ed. T. Adorno. Harper and Row.

Porritt, Jonathan. 1994. The environmental crisis and the green movement. *Utne Reader,* March-April: 60–62.

Porter, Eliot, and David Brower. 1966. *The Place No One Knew: Glen Canyon on the Colorado.* Sierra Club.

Powell, John Wesley. 1878. *Report on the Lands of the Arid Region of the United States* (Harvard University Press, 1962).

Powell, Walter W., and Rebecca Friedkin. 1987. Organizational change in nonprofit organizations. In *The Nonprofit Sector*, ed. W. Powell. Yale University Press.

Pursell, Carroll. 1993. The rise and fall of the appropriate technology movement in the United States, 1965–1985. *Technology and Culture* 34, no. 3: 629–637.

Rabinow, Paul, ed. 1984. *The Foucault Reader*. Pantheon.

Radnitzky, G. 1973. *Contemporary Schools of Metaphysics*, third edition. Regnery.

Ramos, Tarso. 1995. Wise Use in the West: The Case of the Northwest Timber Industry. In *Let the People Judge*, ed. J. Echeverria and R. Eby. Island.

Randall, John Herman, Jr. 1926. *The Making of the Modern Mind* (Columbia University Press, 1976).

Ray, Larry J. 1993. *Rethinking Critical Theory: Emancipation in the Age of Global Social Movements*. Sage.

Rayner, S. 1992. Review of global environmental change: Understanding the human dimensions. *Environment* 34, no. 7: 25–29.

Reese, Robert M. 1908, The spendthrift. *Forestry and Irrigation*, May 1908.

Reiger, John F. 1975. *American Sportsmen and the Origins of Conservation*. Winchester.

Reneau, J., and S. Reneau. 1993 *Records of North American Big Game*, tenth edition. Boone and Crockett Club.

Renn, O., T. Webler, and P. Wiedemann, eds. 1995. *Fairness and Competence in Citizen Participation: Evaluation Models for Environmental Discourse*. Kluwer.

Rheem, Donald L. 1987. Environmental action: A movement comes of age. *Christian Science Monitor*, January 13.

Richards, Ellen. 1912. *Euthenics: The Science of Controllable Environment*. Whitcomb and Barrows.

Richards, Ellen. 1915, *The Art of Right Living*. Whitcomb and Barrows.

Richards, Ellen. 1882. *The Chemistry of Cooking and Cleaning: A Manual for Housekeepers*. Estes & Lauriat.

Richardson, Elmo R. 1962. *The Politics of Conservation: Crusades and Controversies 1897–1913*. University of California Press.

Ricoeur, Paul. 1981. *Hermeneutics and the Human Sciences*. Cambridge University Press.

Riesman, David. 1950, *The Lonely Crowd*. Yale University Press

Rittner, Don. 1992. *Ecolinking*. Peachpit.

Robbins, Roy M. 1962. *Our Landed Heritage: The Public Domain, 1776–1936*. University of Nebraska Press.

Robinson, Marshall. 1993. The Ford Foundation: Sowing the Seeds of a Revolution Environment. *Please Insert Title of Journal* 35, no. 3: 10–41.

Roderick, Rick. 1986. *Habermas and the Foundations of Critical Theory*. Macmillan.

Roe, Emery. 1994 *Narrative Policy Analysis: Theory and Practice*. Duke University Press.

Roelofs, J. 1986. Do foundations set the agenda? From social protest to social service. Presented at Conference of Philanthropy Project, Minneapolis.

Rogers, Marion Lane. 1990. *Acorn Days: The Environmental Defense Fund and How It Grew*. Environmental Defense Fund.

Rohter, Ira. 1993. The U.S. Green Party: Challenges and prospects. Presented at Annual Meeting of American Political Science Association, Washington.

Rootes, C. A. 1983. On the social structural sources of political conflict: An approach from the sociology of knowledge. *Research in Social Movements, Conflict, and Change* 5: 33–54.

Rorty, Amelie O. 1987. Persons as rhetorical categories. *Social Research* 54, no. 1: 55–72.

Rosa, Eugene A. 1997. Cross-national trends in fossil fuel consumption, societal well-being, and carbon releases. In *Environmentally Significant Consumption*, ed. P. Stern et al. National Academy Press.

Rose, Chris. 1993. Beyond the struggle for proof: Factors changing the environmental movement. *Environmental Values* 2: 285–298.

Rosen, George. 1958. *A History of Public Health*. MD Publications.

Rosenberg, Bernard, and David M. White, eds. 1957. *Mass Culture: The Popular Arts in America*. Free Press.

Rossi, Peter H., Howard E. Freeman, and Sonia R. Wright. 1979. *Evaluation: A Systematic Approach*. Sage.

Rothenberg, David. 1992. Does the ecology movement have a philosophy? *Social Policy*, spring: 49–55.

Roussopoulos, Dimitrios Il. 1993. *Political Ecology*. Black Rose Books.

Routley, Richard. 1983. Roles and limits of paradigm in environmental thought and action. In *Environmental Philosophy*, ed. R. Elliot and A. Gare. Penn State Press.

Roy, W. G. 1984. Class conflict and social change in historical perspective. *Annual Review of Sociology* 10: 483–506.

Rucht, Dieter. 1988. Themes, logics and arenas of social movements: A structural approach. *International Social Movement Research* 1: 305–328.

Rucht, Dieter. 1989. Environmental movement organizations in West Germany and France: Structure and interorganizational relations. *International Social Movement Research* 2: 61–94.

Rucht, Dieter. 1991. *Research on Social Movements: The State of the Art in Western Europe and the USA*. Westview.

Rucht, Dieter, and Doug McAdam. 1993. The cross-national diffusion of movement ideas. *Annals of the American Academy of Political and Social Science* 528: 56–74.

Rudig, Wolfgang. 1991. Green party politics around the world. *Environment* 33, no. 8: 6–9.

Ruther, Rosemary Radford. 1975. *Sexism and God-Talk: Toward a Feminist Theology*. Beacon.

Salisbury, S. 1996. Private money, public power: Pew Charitable Trusts develops a hands-on role. *Philadelphia Inquirer*, Sunday, October 13.

Salzman, H., and G. W. Domhoff. 1983. Nonprofit organizations and the corporate community. *Social Science History* 7, no. 2: 205–215.

Sandbach, Francis. 1980. *Environmental Ideology and Policy*. Allanheld, Osmun.

Saussure, Ferdinand de. 1966. *Course in General Linguistics*. McGraw-Hill.

Scarce, Rik. 1990. *Eco-Warriors: Understanding the Radical Environmental Movement*. Noble.

Schaefer, Hart. 1991. How to deal with the Sierra Club. *Wild Earth* 1, no. 1: 17–18.

Schaeffer, Francis A. 1968. *Pollution and the Death of Man: The Christian View of Ecology*. Tyndale House.

Scheffer, Victor B. 1991. *The Shaping of Environmentalism in America*. University of Washington Press.

Schlosberg, David. 1998. *Environmental Justice and the New Pluralism: The Challenge of Difference for Environmentalism*. Oxford University Press.

Schmitt, Peter J. 1969. *Back to Nature: The Arcadian Myth in Urban America*. Johns Hopkins University Press.

Schmitter, Philippe C. 1983. Democratic theory and neocorporatist practice. *Social Research* 50, no. 4: 851–884.

Schmitter, Philippe C., and G. Lehmbruch. 1979. *Trends Toward Corporatist Intermediation*. Sage.

Schnaiberg, Alan. 1980. *The Environment: From Surplus to Scarcity*. Oxford University Press.

Schnaiberg, Alan, and Kenneth Gould. 1994. *Environment and Society: The Enduring Conflict*. St. Martin's Press.

Schneider, J. W. 1985. Social problems theory: The constructionist view. *Annual Review of Sociology* 11: 209–229.

Schudson, M. 1984. Embarrassment and Erving Goffman's idea of human nature. *Theory and Society* 13, no. 5: 633–648.

Schwab, Jim. 1994. *Deeper Shades of Green: The Rise of Blue-Collar and Minority Environmentalism in America*. Sierra Club.

Scott, Alan. 1990. *Ideology and the New Social Movements*. Unwin Hyman.

Scudder, Samuel H. 1884. The Alpine Club of Williamstown, Mass. *Appalachia*, December.

Sessions, George. 1992. Radical Environmentalism in the 90s. In *After Earth Day*, ed. M. Oelschlaeger. University of North Texas Press.

Sessions, George. 1994. Ecocentrism and the Anthropocentric Detour. In *Ecology*, ed. C. Merchant and R. Gottlieb. Humanities.

Shabecoff, Philip. 1993. *A Fierce Green Fire: The American Environmental Movement*. Hill and Wang.

Shaler, Nathaniel Southgate. 1905, *Man and the Earth*. Fox, Duffield.

Short, Brant. 1989. *Ronald Reagan and the Public Lands*. Texas A&M University Press.

Short, Brant. 1991. Earth First! and the rhetoric of moral confrontation. *Communication Studies* 42, no. 2: 172–188.

Shrader-Frechette, K. S., and Earl D. McCoy. 1994. How the tail wags the dog: How value judgements determine ecological science. *Environmental Values* 3: 107–120.

Silverstein, K., A. Cockburn, and J. St. Clair. 1995. How oil money buys liberals. *Counterpunch* 2, no. 19: 3.

Simmel, Georg. 1907. *The Philosophy of Money*. Reprint: Routledge and Kegan Paul. 1978.

Simon, Herbert W. 1990. *The Rhetorical Turn: Invention and Persuasion in the Conduct of Inquiry*. University of Chicago Press.

Slayton, Christa Daryl. 1992. The failure of the United States Greens to root in fertile soil. *Research in Social Movements, Conflicts and Change*, supplement 2: 83–118.

Smith, D. E. 1979. A Sociology for Women. In *The Prism of Sex*, ed. J. Sherman and E. Beck. University of Wisconsin Press.

Smith, James Noel. 1974. The Coming of Age of Environmentalism in American Society. In *Environmental Quality and Social Justice in Urban America*, ed. J. Smith. Conservation Foundation.

Snow, David A., and R. D. Benford. 1988. Ideology, Frame Resonance and Participant Mobilization. In *International Social Movement Research*, ed. B. Klandermans et al. JAI.

Snow, David A., and R. D. Benford. 1992. Master frames and cycles of protest. In *Frontiers in Social Movement Theory*, ed. A. Morris and C. Mueller. Yale University Press.

Snow, David A., and Richard Machalek. 1984. The Sociology of Conversion. *Annual Review of Sociology* 10.

Snow, David A., and S. E. Marshall. 1984. Cultural imperialism, social movements, and the Islamic revival. *Research in Social Movements, Conflict, and Change* 7: 131–152.

Snow, David A., E. Burke Rochford Jr., Steven K. Worden, and Robert D. Benford. 1986. Frame alignment processes, micromobilization, and movement participation. *American Sociological Review* 51, no. 4: 464–481.

Snow, Donald. 1992. *Inside the Environmental Movement: Meeting the Leadership Challenge*. Island.

Snyder, Gary. 1974. *Turtle Island*. New Directions.

Snyder, Gary. 1990. *The Practice of the Wild*. North Point.

Southwest Organizing Project. 1990. Letter to National Environmental Groups, March 16.

Spector, M., and J. I. Kitsuse. 1977. *Constructing Social Problems*. Cummings.

Staggenborg, Suzanne. 1989. Organizational and environmental influences on the development of the pro-choice movement. *Social Forces* 68, no. 1: 204–240.

Stanley, Manfred. 1978. *The Technological Conscience*. University of Chicago Press.

Stapleton, Richard M. 1992. Greed vs. Green. *National Parks Magazine*, November-December.

Stapleton, Richard M. 1993. On the western front. *National Parks Magazine*, January-February.

Steiner, Frederick R. 1990. *Soil Conservation in the United States: Policy and Planning*. Johns Hopkins University Press.

Stern, Paul C. 1991. Learning through conflict: A realistic strategy for risk communication. *Policy Sciences* 24: 99–119.

Stern, Paul C. 1993. A second environmental science: human-environment interactions. *Science* 260: 1897–1899.

Stern, Paul C., and Thomas Dietz. 1994. The value basis of environmental concern. *Journal of Social Issues* 50, no. 3: 65–84.

Stern, Paul C., and H. V. Fineberg, eds. 1996. *Understanding Risk: Informing Decisions in a Democratic Society*. National Academy Press.

Stern, Paul C., T. Dietz, and J. S. Black. 1986. Support for environmental protection: The role of moral norms. *Population and Environment* 8: 204–222.

Stern, Paul C., T. Dietz, and L. Kalof. 1992. Value orientations, gender, and environmental concern. *Environment and Behavior* 25, no. 3: 322–348.

Stern, Paul, T. Dietz, V. Ruttan, R. Socolow, and J. Sweeney, eds. 1997. *Environmentally Significant Consumption: Research Directions*. National Research Council.

Stern, Paul C., O. Young, and D. Druckman. 1992. *Global Environmental Change: Understanding the Human Dimensions*. National Academy Press.

Stewart, Charles J. 1980. A functional approach to the rhetoric of social movements. *Central States Speech Journal* 31, no. 4: 298–305.

Stewart, Charles J., Craig Allen Smith, and Robert E. Denton. 1989. *Persuasion and Social Movements*. Waveland.

Stivers, Richard. 1989. The concealed rhetoric of sociology: Social problem as a symbol of evil. Presented at conference on Writing the Social Text, University of Maryland.

Strasssman, J. D. 1978. *The Limits of Technocratic Politics*. Transaction.

Strydom, Piet. 1993 Collective learning: Habermas's concessions and their theoretical implications. *Philosophy and Social Criticism* 13, no. 3: 265–281.

Swain, George F. 1910. Conservation of Natural Resources: Report of the Conservation Committee of the American Unitarian Association.

Swain, Robert E. 1923. *Atmospheric Pollution by Industrial Wastes.* Columbia University Press.

Szasz, Andrew. 1992. Progress through mischief: The social movement alternative to secondary associations. *Politics and Society* 20, no. 4: 521–528.

Szasz, Andrew. 1994. *EcoPopulism: Toxic Waste and the Movement for Environmental Justice.* University of Minnesota Press.

Takacs, David. 1996. *The Idea of Biodiversity: Philosophies of Paradise.* Johns Hopkins University Press.

Tarr, Joel A. 1984. The search for the ultimate sink: urban air, land and water pollution in historical perspective. *Records of the Columbia Historical Society* 51: 1–29.

Tarr, Joel A. 1985. Historical perspectives on hazardous wastes in the United States. *Waste Management and Research* 3: 95–102.

Taylor, B. P. 1992. *Our Limits Transgressed: Environmental Political Thought in America.* University Press of Kansas.

Taylor, Dorceta E. 1993. Environmentalism and the politics of inclusion. In *Confronting Environmental Racism*, ed. R. Bullard. South End.

Taylor, Serge. 1984. *Making Bureaucracies Think.* Stanford University Press.

Teymur, Necdet. 1982. *Environmental Discourse: A Critical Analysis of "Environmentalism" in Architecture, Planning, Design, Ecology, Social Sciences, and the Media.* ?uestion.

Thompson, E. B. 1963. *The Making of the English Working Class.* Gollancz.

Thompson, J. B. 1984. *Studies in the Theory of Ideology.* University of California Press.

Thoreau, Henry David. 1858. Chesuncook. *Atlantic Monthly*, June: 1–12; July: 224–233; August: 305–317.

Thoreau, Henry David. 1862. *Walking* (Applewood, 1992).

Tilly, Charles. 1978. *From Mobilization to Revolution.* Random House.

Tilly, Charles. 1981. *As Sociology Meets History.* Academic Press.

Tilly, Charles. 1985. Models and realities of popular collective action. *Social Research* 52, no. 4.: 717–748.

Tober, James A. 1989. *Wildlife and the Public Interest: Nonprofit Organizations and Federal Wildlife Policy.* Praeger.

Tokar, Brian. 1995. The wise use backlash: Responding to militant anti-environmentalism. *The Ecologist* 25, no. 4: 150–156.

Torgerson, Douglas. 1995. The uncertain quest for sustainability: Public discourse and the politics of environmentalism. In *Greening Environmental Policy*, ed. F. Fischer and M. Black. St. Martins Press.

Torgerson, Douglas. 1999. *The Promise of Green Politics: Environmentalism and the Public Sphere*. Duke University Press.

Touraine, Alain. 1977. *The Voice and the Eye*. Cambridge University Press.

Trainer, Ted. 1998. Our unsustainable society: Basic causes, interconnections, and solutions. In *The Coming Age of Scarcity*, ed. M. Dobkowski and I. Wallimann. Syracuse University Press.

Trefethen, James B. 1961. *Crusade for Wildlife: Highlights in Conservation Progress*. Boone and Crockett Club.

Trefethen, James B. 1964 *Wildlife Management and Conservation*. Heath.

Tucker, Kenneth H. 1991. How new are the new social movements? *Theory, Culture and Society* 8, no. 2: 75–98.

Turner, Jonathan. 1992. The Promise of Positivism. In *Postmodernism and Social Theory*, ed. S. Seidman and D. Wauner. Blackwell.

UCC (United Church of Christ). 1987. Toxic Wastes and Race in the United States: A National Report on the Racial and Socio-Economic Characteristics of Communities with Hazardous Waste Sites. Commission for Racial Justice

Udall, Stewart L. 1963. *The Quiet Crisis*. Holt, Rinehart and Winston.

U.S. Congress. 1993. House of Representatives, Committee on Energy and Commerce, Subcommittee on Transportation and Hazardous Materials, Hearing on Environmental Justice, November 18.

U.S. Department of Agriculture. 1890. *Report of the Secretary of Agriculture: 1890*. Government Printing Office.

U.S. Department of Agriculture. 1899. *Yearbook of the United States Department of Agriculture: 1898*. Government Printing Office.

U.S. Department of the Interior. 1950. *Conservation Bulletin 39: A Century of Conservation: 1849–1949*. Government Printing Office.

U.S. Department of the Interior. 1959., *Conservation Bulletin 41: Highlights in the History of Forest and Related Natural Resource Conservation*.

U.S. District Court, District of Columbia. 1979. *Health Research Group v Kennedy and Public Citizen* [82 F. R. D. 21 (1979)].

U.S. Executive Office. 1909. Proceedings of a Conference of Governors in the White House, May 13–15, 1908.

U.S. Executive Office. 1963. President's Science Advisory Committee, Life Sciences Panel Report on the Use of Pesticides, Food Drug Cosmetic Law Reports, no. 10, part 1. Commerce Clearing House.

U.S. Executive Office. 1994. Executive Order no. 12898, Federal Actions to Address Environmental Justice in Minority Populations and Low-Income Populations.

U.S. General Accounting Office. 1983. Siting of Hazardous Waste Landfills and Their Correlation with Racial and Economic Status of Surrounding Communities.

U.S. Great Plains Committee. 1936. *The Future of the Great Plains*. Government Printing Office.

U.S. National Resources Board. 1934. *A Report on National Planning and Public Works in Relation to Natural Resources and Including Land Use and Water Resources with Findings and Recommendations*, Government Printing Office.

U.S. Special Committee on Conservation of Wildlife Resources. 1936, *Wildlife Restoration and Conservation: Proceedings of the North American Wildlife and Natural Resources Conference, February 3–7, 1936*, Government Printing Office.

U.S. Supreme Court. 1972. *Sierra Club v. Morton* [405 US 727].

U.S. Supreme Court. 1977. *Hunt v. Washington Apple Advertising Commission* [432 US 76-63].

Ufford, Walter Shepard. 1897. Fresh Air Charity in the United States. Ph.D. dissertation, Columbia University.

Van Hise, Charles R. 1921. History of the conservation movement. In *Readings in Resource Management and Conservation*, ed. I. Burton and R. W. Kates. University of Chicago Press.

Veblen, Thorstein. 1899. *The Theory of the Leisure Class*. Reprint: Macmillan. 1979.

Vitousek, P., H. Mooney, J. Lubchenco, and J. Melillo. 1997. Human domination of Earth's ecosystems. *Science* 277: 494–499.

Vogel, David. 1987. A big agenda. *Wilson Quarterly*, autumn: 51–68.

Vogel, Steven. 1997. Habermas and the ethics of nature. In *The Ecological Community*, ed. R. Gottlieb. Routledge.

Vogt, William. 1948. *Road to Survival*. W. Sloane Associates.

Walker, Jack L. 1983. The origins and maintenance of interest groups in America. *American Political Science Review* 77: 390–406.

Walker, Jack L. 1991. *Mobilizing Interest Groups in America: Patrons, Professions, and Social Movements*. University of Michigan Press.

WCED (World Commission on Environment and Development). 1987. *Our Common Future*. Oxford University Press.

Weber, Max. 1904. *The Protestant Ethic and the Spirit of Capitalism*. Reprint: Scribner, 1958.

Weber, Max. 1918. Politics as a Vocation. Reprinted in *From Max Weber: Essays in Sociology*, ed. H. Gerth and C. Mills (Oxford University Press, 1978).

Webler, Thomas. 1992. Habermas put into practice: A democratic discourse for environmental problem solving. Presented at Sixth Annual Meeting of Society for Human Ecology, Snowbird, Utah.

Webler, Thomas, Hans G. Kastenholz, and Ortwin Renn. 1994. Can Public Participation in Impact Assessment Enable Social Learning? Presented at Annual Meeting of Society for Human Ecology.

Weisberg, Barry. 1971. *Beyond Repair: The Ecology of Capitalism*. Beacon.

Whaley, Rick, and Walter Bresette. 1994. *Walleye Warriors: An Effective Alliance against Racism and for the Earth*. New Society.

White, Lynn. 1967. The historical roots of our ecologic crisis. *Science* 155, no. 3767: 1203–1207.

Whitebook, Joel. 1979. The problem of nature in Habermas. *Telos* 40: 41–69.

Whyte, William H., Jr. 1956. *The Organization Man*. Simon and Schuster.

Williams, Bruce A., and Albert R. Matheny. 1995. *Democracy, Dialogue, and Environmental Disputes: The Contested Languages of Social Regulation*. Yale University Press.

Willoughby, Kelvin W. 1990. *Technology Choice: A Critique of the Appropriate Technology Movement*. Westview.

Wilson, Frank L. 1990. Neo-corporatism and the Rise of New Social Movements. In *Challenging the Political Order*, ed. R. Dalton and M. Kuechler. Oxford University Press.

Wing, William G. 1973. *Philanthropy and the Environment*. Conservation Foundation.

Wolf, Hazel. 1994. The founding mothers of environmentalism. *Earth Island Journal*, winter 1993–1994: 36–37.

Worster, Donald., ed. 1973. *American Environmentalism: The Formative Period, 1860–1915*. Wiley.

Worster, Donald. 1977. *Nature's Economy: A History of Ecological Ideas*. Cambridge University Press.

Worster, Donald. 1994. Nature and the disorder of history, *Environmental History Review* 18, no. 2: 1–15.

Worster, Donald. 1985a. Comment: Hays and Nash. In *Environmental History*, ed. K. Bailes. University Press of America.

Worster, Donald. 1985b. *Nature's Economy: A History of Ecological Ideas*. Cambridge University Press.

Wright, Mabel Osgood. 1903. *People of the Whirlpool*. Macmillan.

Wright, Mabel Osgood. 1909. *Poppea of the Post Office*. Macmillan.

Yearley, S. 1994. Social movements and environmental change. In *Social Theory and the Global Environment*, ed. T. Benton and M. Redclift. Routledge.

Ylvisaker, Paul N. 1987. Foundations and nonprofit organizations. In *The Nonprofit Sector*, ed. W. Powell. Yale University Press.

Young, John. 1990. *Sustaining the Earth: The Story of the Environmental Movement*. Harvard University Press.

Zald, Mayer N. 1992. Looking backward to look forward: Reflections on the past and future of the resource mobilization research program. In *Frontiers in Social Movement Theory*, ed. A. Morris and C. Mueller. Yale University Press.

Zald, Mayer N., and Roberta Ash Garner. 1987. Social movement organizations: Growth, decay, and change. In *Social Movements in an Organizational Society*, ed. M. Zald and D. McCarthy. Transaction.

Zald, Mayer N., and John D. McCarthy. 1987. *Social Movements in an Organizational Society*. Transaction.

Zarefsky, David. 1980. A skeptical view of movement studies. *Central States Speech Journal* 31, no. 4: 245–254.

Zeitlin, Maurice. 1980. *Classes, Class Conflict, and the State: Empirical Studies in Class Analysis*. Winthrop.

Zimmerman, Michael E. 1990. Deep Ecology and Ecofeminism: The Emerging Dialogue. In *Reweaving the World*, ed. I. Diamond and G. Orenstein. Sierra Club.

Zimmerman, Michael E. 1994. *Contesting Earth's Future: Radical Ecology and Postmodernity*, Berkeley. University of California Press.

Zimmerman, Michael E. 1997. Ecofascism: A threat to American environmentalism? In *The Ecological Community*, ed. R. Gottlieb. Routledge.

Index